Models That Work

Case Studies
in
Effective Undergraduate Mathematic

Models That Work

Case Studies
in
Effective Undergraduate Mathematics Programs

An MAA project funded by the National Science Foundation

Alan C. Tucker, Editor
SUNY at Stony Brook

Published by
The Mathematical Association of America

This project was supported, in part, by the National Science Foundation.
Opinions expressed in this report are those of the authors and not necessarily
those of the foundation.

MAA Notes and Reports Series

The MAA Notes and Reports Series, started in 1982, addresses a broad range of topics and themes of interest to all who are involved with undergraduate mathematics. The volumes in this series are readable, informative, and useful, and help the mathematical community keep up with developments of importance to mathematics.

MAA Notes

1. Problem Solving in the Mathematics Curriculum, *Committee on the Teaching of Undergraduate Mathematics*, a subcommittee of the Committee on the Undergraduate Program in Mathematics, *Alan H. Schoenfeld*, Editor

2. Recommendations on the Mathematical Preparation of Teachers, *Committee on the Undergraduate Program in Mathematics, Panel on Teacher Training*.

3. Undergraduate Mathematics Education in the People's Republic of China, *Lynn A. Steen*, Editor.

5. American Perspectives on the Fifth International Congress on Mathematical Education, *Warren Page*, Editor.

6. Toward a Lean and Lively Calculus, *Ronald G. Douglas*, Editor.

8. Calculus for a New Century, *Lynn A. Steen*, Editor.

9. Computers and Mathematics: The Use of Computers in Undergraduate Instruction, *Committee on Computers in Mathematics Education, D. A. Smith, G. J. Porter, L. C. Leinbach, and R. H. Wenger*, Editors.

10. Guidelines for the Continuing Mathematical Education of Teachers, *Committee on the Mathematical Education of Teachers*.

11. Keys to Improved Instruction by Teaching Assistants and Part-Time Instructors, *Committee on Teaching Assistants and Part-Time Instructors, Bettye Anne Case*, Editor.

13. Reshaping College Mathematics, *Committee on the Undergraduate Program in Mathematics, Lynn A. Steen*, Editor.

14. Mathematical Writing, by *Donald E. Knuth, Tracy Larrabee, and Paul M. Roberts*.

15. Discrete Mathematics in the First Two Years, *Anthony Ralston*, Editor.

16. Using Writing to Teach Mathematics, *Andrew Sterrett*, Editor.

17. Priming the Calculus Pump: Innovations and Resources, *Committee on Calculus Reform and the First Two Years*, a subcomittee of the Committee on the Undergraduate Program in Mathematics, *Thomas W. Tucker*, Editor.

18. Models for Undergraduate Research in Mathematics, *Lester Senechal*, Editor.

19. Visualization in Teaching and Learning Mathematics, *Committee on Computers in Mathematics Education, Steve Cunningham and Walter S. Zimmermann*, Editors.

20. The Laboratory Approach to Teaching Calculus, *L. Carl Leinbach et al.*, Editors.

21. Perspectives on Contemporary Statistics, *David C. Hoaglin and David S. Moore*, Editors.

22. Heeding the Call for Change: Suggestions for Curricular Action, *Lynn A. Steen*, Editor.

23. Statistical Abstract of Undergraduate Programs in the Mathematical Sciences and Computer Science in the United States: 1990–91 CBMS Survey, *Donald J. Albers, Don O. Loftsgaarden, Donald C. Rung, and Ann E. Watkins*.

24. Symbolic Computation in Undergraduate Mathematics Education, *Zaven A. Karian*, Editor.

25. The Concept of Function: Aspects of Epistemology and Pedagogy, *Guershon Harel and Ed Dubinsky*, Editors.

26. Statistics for the Twenty-First Century, *Florence and Sheldon Gordon*, Editors.

27. Resources for Calculus Collection, Volume 1: Learning by Discovery: A Lab Manual for Calculus, *Anita E. Solow*, Editor.

28. Resources for Calculus Collection, Volume 2: Calculus Problems for a New Century, *Robert Fraga*, Editor.

29. Resources for Calculus Collection, Volume 3: Applications of Calculus, *Philip Straffin*, Editor.

30. Resources for Calculus Collection, Volume 4: Problems for Student Investigation, *Michael B. Jackson and John R. Ramsay*, Editors.

31. Resources for Calculus Collection, Volume 5: Readings for Calculus, *Underwood Dudley*, Editor.

32. Essays in Humanistic Mathematics, *Alvin White*, Editor.

33. Research Issues in Undergraduate Mathematics Learning: Preliminary Analyses and Results, *James J. Kaput and Ed Dubinsky*, Editors.

34. In Eves' Circles, *Joby Milo Anthony*, Editor.

35. You're the Professor, What Next? Ideas and Resources for Preparing College Teachers, *The Committee on Preparation for College Teaching, Bettye Anne Case*, Editor.

36. Preparing for a New Calculus: Conference Proceedings, *Anita E. Solow*, Editor.

37. Practical Guide to Cooperative Learning, *Nancy Hagelgans, Barbara E. Reynolds, SDS, Keith Schwingendorf, Draga Vidakovic, Ed Dubinsky, Mazen Shahin, G. Joseph Wimbish, Jr.*

38. Models That Work: Case Studies in Effective Undergraduate Mathematics Programs, *Alan C. Tucker*, Editor.

MAA Reports

1. A Curriculum in Flux: Mathematics at Two-Year Colleges, *Subcommittee on Mathematics Curriculum at Two-Year Colleges*, a joint committee of the MAA and the American Mathematical Association of Two-Year Colleges, *Ronald M. Davis*, Editor.

2. A Source Book for College Mathematics Teaching, *Committee on the Teaching of Undergraduate Mathematics, Alan H. Schoenfeld*, Editor.

3. A Call for Change: Recommendations for the Mathematical Preparation of Teachers of Mathematics, *Committee on the Mathematical Education of Teachers, James R. C. Leitzel*, Editor.

4. Library Recommendations for Undergraduate Mathematics, *CUPM ad hoc Subcommittee, Lynn A. Steen*, Editor.

5. Two-Year College Mathematics Library Recommendations, *CUPM ad hoc Subcommittee, Lynn A. Steen*, Editor.

6. Assessing Calculus Reform Efforts: A Report to the Community, *Alan C. Tucker and James R. C. Leitzel*, Editors.

To order any of these volumes call
800-331-1MAA

Preface

This case studies project reflects the current philosophy of MAA education activities: to assist members in their instructional efforts rather than to attempt to offer guidance to them. Guidance was implicit in the past CUPM course syllabus recommendations. This project report is meant as a resource for faculty seeking to improve their undergraduate programs. It summarizes effective practices at a set of mathematics departments that are excelling at

(i) attracting and training large numbers of mathematics majors, or

(ii) preparing students to pursue advanced study in mathematics, or

(iii) preparing future school mathematics teachers, or

(iv) attracting and training underrepresented groups in mathematics.

The site visits revealed that there is no single key to a successful undergraduate program in mathematics. Almost any approach can be made to work in almost any institutional context if a substantial number of the mathematical faculty care deeply about undergraduate education, create an atmosphere among faculty and students that the study of mathematics is important and rewarding, and maintain close interactions with their students.

The advisory committee consisted of

Prof. Linda Boyd, Chair, MAA Two-Year College Committee; Mathematics Department, DeKalb Community College.

Dr. Susan Forman, MAA First Vice-President 1992–93; MSEB Director of College-University Programs 1993–95.

Dr. William Hawkins, Director of MAA SUMMA Project; on leave from Howard University.

Prof. Deborah Hughes-Hallett, Mathematics Department, Harvard University.

Prof. Harvey Keynes, co-Director of the Mathematics and Education Reform Network and Chair, MSEB College-University Panel; Mathematics Department, University of Minnesota.

Prof. James Leitzel, Chair, MAA Committee on the Undergraduate Program in Mathematics; Professor of Mathematics, University of Nebraska-Lincoln.

Prof. William Lucas, former Chair, MAA Committee on the Undergraduate Program in Mathematics; Mathematics Department, Claremont Graduate School.

Prof. David Lutzer, Chair, MAA Committee on Advising; Dean of Arts and Sciences, College of William and Mary.

Prof. Ray Shiflett, past MSEB Director of College-University Programs; Mathematics Department, California State Polytechnical University-Pomona.

Prof. Ivar Stakgold, former SIAM President; Associate Executive Director for AMS Washington Office, on leave from Department of Mathematics Sciences, University of Delaware.

Dr. Marcia Sward, MAA Executive Director.

Prof. Irvin Vance, Mathematics Department, Michigan State University.

The advisory board members developed the set of questions that guided the site visits, undertook the project site visits, and aggressively oversaw the writing of the project report.

A report such as this inevitably contains some conclusions and opinions that are not fully substantiated with references or persuasive arguments. These are subjective judgments that represent the collective beliefs of the project members. With these as with all statements in this report, it is up to individual readers to decide what is useful to them. The project members gratefully acknowledge the support of the National Science Foundation for funding this effort, grant DUE 9155957.

Alan Tucker, Project Director
August, 1995

Contents

Preface vii

I.   Introduction 1

II.   Common Features of Effective Programs 3

III.   Attracting Students to Study Mathematics: In-Class Experiences 4

IV.   Attracting Students to Study Mathematics: Out-of-Class Experiences 11

V.   Attracting High School Students to Study Mathematics 14

VI.   Organization of the Mathematics Major and Supporting Activities 17

VII.   Effective Programs for Preparing Students for Advanced Study 23

VIII.   Effective Programs for Preservice Preparation of School Mathematics Teachers 26

IX.   Effective Programs for Underrepresented Groups in Mathematics 29

X.   Concluding Remarks 32

Appendices—Site Visit Reports

    Lebanon Valley College 33

    Miami University of Ohio 38

    Mount Holyoke College 42

    Saint Olaf College 47

    Seattle Central Community College 52

    Southern University 54

    Spelman College 59

    University of Chicago 63

    University of Michigan 70

    University of New Hampshire 76

The major program in mathematics is based on the premise that the study of pure mathematics can be undertaken successfully by a large number of students if they are provided with a supportive environment including: careful and considerate teaching by a well-trained and dedicated faculty, continual encouragement, successful (student) role models, enough success to develop self-esteem, enough time to develop intellectually, recognition of their achievement, and the belief that the study is a worthwhile endeavor. We are dedicated to providing this supportive environment.

from Department Mission Statement,
SUNY-College at Potsdam Mathematics Department

I. Introduction

This report complements MAA curriculum recommendations about the undergraduate program in mathematics with a case studies approach describing effective practices in undergraduate mathematics programs. The primary focus of this case studies project is not course content but rather more general issues in the undergraduate enterprise such as effective instruction, advising, tailoring the curriculum to students' needs, and interactions between students and mathematics faculty.

The heart of this project is a set of site visits to ten mathematics departments with undergraduate programs that are considered effective in one or more of the following categories:

- attracting and preparing large numbers of mathematics majors, or

- preparing students to pursue advanced study in mathematics, or

- preparing future school mathematics teachers, or

- attracting and preparing underrepresented groups in mathematics.

They spanned the spectrum from two-year to doctoral institutions. This report also draws on the experiences of the advisory board with other highly effective mathematics programs.

The objective of this project is to share with the mathematics community activities and attitudes that have been quite effective at some departments. The project advisory committee has been very conscious of not wanting to put a small number of mathematics programs up on a pedestal. Not only are there many very effective programs that did not receive site visits, but also the project members know of many programs that, given local conditions, are doing an impressive job even though they are not excelling at one of the four goals listed above. The MAA gopher on the World Wide Web has a growing electronic database, Celebrating Progress in Mathematics Education, about such programs.

It is probably not surprising that the effective undergraduate programs encountered in this project do many things well—many mathematics majors, a substantial number of majors going to graduate school, good participation of underrepresented groups, and good teacher training. What was a bit unexpected was the common attitude in effective programs that the faculty are not satisfied with the current program. They are constantly trying innovations and looking for improvement. Quoting one site visit report: "An overriding impression of the department is that it is never static. There is always discussion of ways to improve course content and instruction." Indeed, there was often a tension between departmental mandates for consistency among different offerings of a course and the urge of individual faculty to tinker.

The site visits demonstrated that there can be a great diversity of instructional and curricular approaches in effective undergraduate program in mathematics. Indeed, diversity of approaches within the mathematics department was a common hallmark of the programs visited. Many approaches can be made to work in almost any institutional context.

This report describes:

(i) general attitudes and strategies as well as particular activities that are effective; and

(ii) ways to create and sustain an environment that will foster such attitudes and activities.

In this report, the term mathematics major is used in an inclusive sense to include majors in mathematics, mathematical sciences, applied mathematics, statistics, actuarial science and the like.

We conclude this introductory section with a brief historical summary of the changing objectives of the undergraduate program in mathematics. The typical mathematics major in 1950 was rigorous in a classical sense of extensive technical facility and complex problem-solving, starting with a year of college algebra and analytical geometry, but it contained little modern analysis or abstract algebra. Internal rethinking and external pressures led in the 1950's and 1960's to a modernization and streamlining of virtually every aspect of the undergraduate curriculum in mathematics. The streamlining to get engineers into and out of calculus sooner, the modernization to prepare mathematics majors better for graduate study (to address the shortage of mathematics PhDs to train the engineers and serve the expanding college population). This modernization involved an increasing emphasis on theory.

In the 1970's, enrollments in upper division theoretical courses such as number theory, abstract algebra and logic collapsed (down over 50% from 1970 to 1975, according to CBMS surveys). The response, typified by the 1981 CUPM *Recommendations for a General Mathematical Sciences Program,* was a more inclusive view of the mathematics major that sought to prepare students for a variety of different mathematical sciences careers. Among the larger numbers of students attracted to the major by BS-level and MS-level professional opportunities, some became motivated to seek advanced training in mathematics. As a consequence, many programs with this inclusive approach actually saw an increase in the number of students going on to doctoral study in mathematics.

II. Common Features of Effective Programs

There are three states of mind that underlie faculty attitudes in effective programs.

- Respecting students, and in particular, teaching for the students one has, not the students one wished one had.

- Caring about the students academic and general welfare.

- Enjoying one's career as a collegiate educator.

These states of mind are held by not just a small core of dedicated faculty in effective programs, but by most or all faculty. A few faculty can make a large difference, but the truly effective programs require a departmental culture that supports these states of mind.

Not surprisingly, when faculty demonstrate respect[1] for their students and enjoyment in their instruction, students respond in kind. Related to respect is an *expectation by faculty that their students will be successful*. The effective programs create an environment of positive mutual re-enforcement. A common theme of effective programs is the existence of a variety of mechanisms for interactions between faculty and students outside of class, both in one-on-one settings and in social groups. Effective programs also foster extensive discussion among students inside and outside class and among faculty. The following quote from the Southern University site visit report illustrates this point: "In one calculus class, the students were actively involved during the entire period. Informal groupings of students helped each other while the teacher presented problems and helped groups when they called upon her. These classroom experiences carry over into informal study sessions in the mathematics lab and dormitories."

Such interactions serve a variety of disparate objectives, all valuable. Foremost, they engage students extensively in mathematical discussions. They foster a professional sense of community and a personal sense of friendship and interest in each other. They provide a framework for feedback to faculty about on-going and new instructional activities. There are two other themes in effective programs that will arise repeatedly in this report:

- Curriculum geared towards the needs of the students, not the values of the faculty.

- An interest in using a variety of pedagogical and learning approaches.

---

[1] There are many ways to try to explain what the term "respect for their students" means. In the classroom, respect is characterized by behavior such as always welcoming student questions and giving constructive interpretations to these questions. It means wanting to get to know one's students so as to anticipate, and accommodate, weaknesses they may have in learning certain types of material.

III. Attracting Students to Study Mathematics: In-Class Experiences

There are two principal reasons why students major in a particular subject in college:

They enjoy studying the subject; or

The study of the subject is thought to lead to attractive career opportunities.

High enrollment programs are effective in attracting majors for both reasons, but some focus more directly on one or the other of these aspects. For example, the University of Chicago is unabashed in its focus on the love of mathematics for its own sake, whereas Vanderbilt University has a very popular mathematical sciences program aimed at professional students, e.g., as a double major with engineering.

Mathematics programs in the 1960's were in a position to be demanding about who could become a mathematics major. Back then, 5% of the incoming college students were interested in mathematics and 2% of the graduates were in mathematics. Now, it is more a matter of attracting students to mathematics. Recent NSF surveys indicate that about 0.7% of entering freshmen intend to become mathematics majors (the number has been quite steady since the 1970's). Yet about 1.8% are graduating from college as mathematics majors. Given an inevitable attrition among those initially planning on a mathematics major, about three-quarters of the students who eventually earn Bachelor's degrees in mathematics enter college planning on another major.

At many institutions today, students are slow to commit themselves to any major or they switch majors often. In such cases, the issue is not so much about getting students to choose a mathematics major as about motivating students to continue taking mathematics courses after calculus. At SUNY-Stony Brook's popular Applied Mathematics major, the majority of students declare the major in the second half of their junior year or later. In other institutions, the choice of additional mathematics courses frequently leads instead to a second major in mathematics. A large fraction of the mathematics majors at some programs with high percent-ages of mathematics graduates, such as Potsdam and Vanderbilt, have double or joint majors. The St. Olaf mathematics major was originally designed to appeal to students as a second major added to a primary-interest major, but its success has led to it being the sole major of most of its students.

The site visits validated the obvious proposition that the best inducement for beginning students to take another mathematics course is to have an excellent teacher in their current course. For this reason, a strong case can be made that a less than excellent teacher can do more damage in Calculus I than in an advanced course. All programs visited gave considerable attention to having good teachers in freshman calculus. If TAs were used, as was the case at Chicago and Michigan, they were given very extensive preparation before being allowed to teach.

Another important factor in successful mathematics programs is interaction among students. These interactions attract new majors as well as motivate and raise the morale of current majors. Quoting from one site report: "Students just beginning their studies see actively engaged upper-class students and mingle with them before and after classes. . . . This critical mass of actively engaged undergraduate students draws more majors into the program."

The following question arises frequently in designing freshman calculus courses and related introductory mathematics courses that may lead to the major: How much attention should be given to students who enter college knowing they want to be mathematics majors? These students typically enjoy doing mathematics, whether it leads to useful careers or not. One does not want to take them for granted and aim one's presentation solely at recruiting 'converts' to mathematics. One also does not want to speak only to the converted, if for no other reason than there are not many such students. Ideally, one seeks to offer something for each kind of student. Effective programs engage all students and motivate them to enjoy learning mathematics and to appreciate its uses and its connections to other disciplines.

1   Classroom Experiences with Instructors

Most faculty in the programs visited approach all courses with a primary focus on the general mathematical experience rather than the particulars of the individual subject. In every class, they try to motivate students to learn and to be interested in mathematics. The particular course syllabus is a context for achieving these broad goals.

A strength of programs that attract large numbers of students into mathematics is the ability of instructors to instill a sense of achievement in students about their learning in every class. At many such institutions, students say that teachers motivate them to realize their potential to excel in, and enjoy, mathematics. The faculty involve the students in the classroom learning by drawing out ideas from students through leading questions and having students come to the board to present solutions to problems. Teachers find ways to make student participation in class a confidence-building experience, not a scary or embarrassing experience. If an instructor appears unsympathetic in answering a slower student's 'dumb' question, all students generally share the instructor's implicit reproach.

Interviews with students repeatedly elicited a feeling from them that their faculty understand what interests them about mathematics— "the faculty are on the same wavelength as us." Students felt that faculty believed in them— expected them to be able to master the mathematics they were studying. The attitudes of instructors towards their students are particularly important for women and underrepresented groups who may feel less comfortable about their mathematics skills and more sensitive to critical behavior, such as showing impatience at a student's question or cutting off a student in the middle of a question and restating the way he/she was posing the question. The mutual respect of students and faculty at the programs visited is felt to be a critical factor in most of these programs higher percentage of women and minority mathematics majors than mathematics programs at peer institutions. As an aside, we note the following well-documented difference between men and women: when a student does poorly on a challenging assignment or test, a male will frequently blame the teacher or bad luck, while a female will frequently blame herself.

Students interviewed during the site visits told the visitors that their faculty set reasonably high standards for them, and then motivate and help them to work hard to meet these standards. Given the wide range of ability of students in most of these programs, it is impressive that virtually all students can feel properly challenged in the same courses.

Much of the preceding discussion can be summed up as follows: the teachers in effective programs are doing good coaching.

2   Placement

The programs visited generally had very carefully designed placement programs. Much faculty effort goes into the tests, and in return, the tests have a high level of credibility with mathematics faculty. Some faculty in other departments used the tests for their own course placement. Chicago, St. Olaf and Southern had particularly strong placement programs.

The placement tests were usually mandatory and were discussed with the student afterwards (several mathematics faculty were always available after the testing to talk to students pleading special circumstances). The departments were wary of students who had "taken" lots of mathematics in high school but knew very little. An example of the thoroughness of such testing is the Southern University program which starts all students with an Arithmetic and Skills test; those that do well take an Advanced Algebra placement test; and those that do well on this get to take a Calculus Readiness test.

Because of the good placement, when the faculty walk into a calculus class they can be confident that everyone in the class is capable of doing well in the course. The faculty are very concerned with letting an eager student get into a course where the student might do poorly and stop taking mathematics forever. The greatest problem seems to be posed by students who had some calculus in high school but not enough to place out of the first semester of calculus. There have been articles about such students and even a special MAA committee to study them. Its report, "Report of the CUPM Panel on Calculus Articulation: Problems in the Transition for High School Calculus—College Calculus," appeared in the October 1987 issue of the *American Mathematical Monthly*. Miami University of Ohio, like several other institutions, has a special calculus course for such students.

The University of Arizona has practice placement tests and adaptive review tutorials on computer disks that are sent to all students the summer before they enroll. Many introductory Arizona mathematics classes have the novel feature of a test in the first two weeks on prerequisite material that will count for 20% of the course grade (students get their scores on this test in time to drop the course if they do badly).

3   Experiences in Learning Calculus

At many institutions, particularly public ones, most future mathematics majors still begin their freshman year with a (single-variable) calculus course. Traditional American calculus courses are based on a syllabus developed in the 1950's to serve the needs of freshman engineering and science majors who were simultaneously taking introductory physics. The focus of traditional courses was techniques of calculus. Honors calculus courses are discussed in the next section.

At the institutions visited, the objective of their calculus courses is to provide a stimulating, meaningful experience in mathematics that also teaches concepts and techniques that will be needed by many students for subsequent coursework in science and engineering. Typically the mathematics involves some theory—taught at varying levels of rigor— and it usually includes a good dose of applications and mathematical modeling. There is an effective and thoughtful balance between problem-solving and mathematical theory, a balance that also highlights the interplay between the two. The instruction is constantly raising questions that require students to grow in their mathematical awareness.

Faculty in the programs visited consciously did little to try to recruit mathematics majors in calculus classes, except to announce informational and social events for prospective mathematics majors and mention that students interested in a particular topic might be interested in certain advanced mathematics courses which developed the topic further. The site report of Spelman is typical "[some students] expect to major in something else but a conversation with their calculus teacher convinced them to major in math. Interestingly enough, the faculty resist— even bristle at—the idea that they recruited students, but simply say that they are making sure the student recognized her potential. This is exactly what they are doing and it has a powerful effect on students."

At the programs visited, the mainstream calculus course is not thought of as a service course. Most mathematics faculty consider calculus to be the most important teaching assignment in the department. One site report stated, "It is in the first semester calculus class that a great deal of attention is showered on students."

A critical objective of calculus courses in the programs visited is to develop each student's sense of self-confidence in doing mathematics. The two-year college mathematics programs visited put particular attention on this theme. At two-year colleges, calculus is near the end of the line, not the beginning of the line, and thus calculus students get confidence-building special attention and encouragement from faculty, just as four-year college faculty give special attention to senior mathematics majors.

Further, two-year college faculty want their calculus students to be successful when they transfer to four-year colleges and universities to complete their bachelor's studies. They view their students as having to work harder to prove themselves (and their college) at their transfer institution. For example, the ability of Seattle Central Community College students to perform well at the University of Washington– and their expectation that they will perform well– has helped increase the pipeline of students taking calculus at Seattle Central and transferring to Washington into heavily quantitative disciplines.

All but one of the programs visited taught calculus in small classes. The one with large lectures, University of New Hampshire, had very effective group learning activities occurring in smaller problem sections. The University of Michigan has made a major push, with substantial support from their administration, to reduce the size of calculus sections from 35 to 24, many taught by faculty (a reform calculus text and group learning activities are part of this initiative).

These programs tended to minimize the number of different tracks of beginning calculus. At institutions as diverse as Spelman College and Miami University of Ohio there was just one calculus sequence (Miami also has an honors calculus sequence and the special course mentioned above for students with some high school calculus). To make this work requires considerable creativity and flexibility. On the other hand, this strategy emphasizes that calculus is about mathematics, not about learning skills needed for a particular major.

4   Honors Courses in Calculus

Honors courses, developed in the 1960's, tended to be high-powered, theoretical courses designed for students who had shown special ability in mathematics. A good number of students in these honors courses went on to become mathematics majors and then continued on to graduate study in mathematics, or at least to take upper-level mathematics courses. In recent years a decreasing number of students in honors calculus courses have become mathematics majors. Indeed, some honors classes are characterized by high dropout rates with fewer than half the original enrollees finishing the full year. Thus the teaching in these courses may unintentionally have more negative than positive impact on recruiting mathematics majors.

Most of the mathematics programs visited have an honors calculus sequence. While these courses are designed for mathematics majors, faculty frequently did not view the honors sequence as the prime source for recruiting majors. An exception was the University of Chicago, which runs a mathematics program little changed from the 1960's. Chicago views its honors calculus course as the primary source of majors. Chicago has high standards for placing out of first-year calculus, and thus many students with AP credit still have to take first-year calculus, a number of whom choose the calculus honors course. The other programs cast a wider net in looking for potential majors.

The enrollment in freshman honors courses has been decreasing because many incoming students interested in a mathematics major have Advanced Placement calculus nowadays and go into sophomore mathematics courses (or into calculus II—placing them out of step with the freshman honors course). There was a feeling at many programs visited that the transition from a high school single-variable calculus course (with 180 course meetings in a year as opposed to 55 to 90 a year in college) to a collegiate multivariable calculus or linear algebra course was challenging enough for freshmen.

Only St. Olaf and the university programs visited, Chicago and Michigan, have a sophomore honors sequence. At most of the four-year college programs visited, sophomore courses, such as linear algebra and differential equations, were populated mainly by mathematics majors, and so special sections for mathematics majors were unnecessary.

5 Impact of Calculus Reform

Most of the emphasis in calculus reform has been on non-content issues, such as how material is taught with a goal of developing better conceptual understanding. The programs visited already knew that the process of learning calculus was at least as important as the details of what was being learned.

It is the satisfying learning experience created by the professors that seems to be the dominant factor in explaining the positive impact of calculus courses at these mathematics programs. Initial reports about calculus reform in general, as documented in the 1995 MAA report, *Assessing Calculus Reform Efforts (ACRE),* found students being more active learners. They also claimed that reformed courses produced an increase in students' confidence in their mathematical ability. The ACRE report found large numbers of faculty reported feeling that teaching a reform calculus course was a much more satisfying experience for them. They became more active teachers as they engaged the students in becoming active learners—a situation typical at the programs visited. Consistent with the increased interest in mathematics at the programs visited, preliminary information in the ACRE Report found that at many departments trying calculus reform there are more students taking advanced mathematics courses. Institutions surveyed by the ACRE study where reform has been in place a couple of years frequently reported some increase in the number of mathematics majors.

The mathematics programs visited in this project were typically using reformed calculus materials in at least some sections, although a few showed little interest in the reform movement. Those actively implementing calculus reform were mostly using reform texts (just as nationally, the most commonly used reform text was the Harvard Consortium text); a few were using traditional texts supplemented by group learning, extensive writing, computer algebra systems, and open-ended projects. These reform efforts included extensive applications and mathematical modeling to demonstrate the power and usefulness of calculus.

Most of the programs visited do not rely heavily on extensive technology (graphing calculators, yes; computers, less commonly), carefully planned cooperative learning or open-ended projects, although a good number of their faculty use one or more of these approaches in some of their courses. As previously noted, these faculty are constantly trying different pedagogical strategies in their classes.

The programs visited that professed no interest in calculus reform have long had some of the pedagogical components of the reform movement in their classrooms. One such program chair said their department was wedded to old-fashioned teaching methods, but when pressed the chair said that a good one-hour lecture should have 5–10 minutes of review or background motivation, 15–20 minutes presenting new material, 15–20 minutes of problem-solving by students (collaboration encouraged) and 15–20 minutes of problems on the board done jointly by teacher and students.

6 Experiences in Mathematics Preceding Calculus

In conversations with students during this project's site visits, there were several students who attributed their interest in majoring in mathematics to a very positive experience in a precalculus course. These students—often older, returning students or students coming from schools with less demanding academic standards—were mathematically able but started college with mathematical deficiencies which forced them into precalculus or remedial courses.

There are a number of mathematics programs where precalculus and first-semester calculus have been combined into an integrated two-semester sequence. Mt. Holyoke has such a sequence that uses a special set of 'workshop' materials to provide the modeling and conceptualizing experiences typical of calculus reform at a precalculus level. St. Olaf is very happy with the results of its new integrated two-semester Calculus with Algebra sequence. A FIPSE-funded conference, "The Integration of Precalculus with Calculus," held at Moravian College in June 1993 sought to share and disseminate results about successful precalculus-calculus efforts at Moravian College and elsewhere. (For more information about presentations at this conference, interested readers should contact Prof. Doris Schattschneider of the Moravian College Mathematics Department.)

There are several NSF-sponsored precalculus reform projects currently in various stages of development, and a variety of innovative precalculus texts will be appearing in the near future. The precalculus reform efforts are most active at two-year colleges. The first reformed precalculus text, *Contemporary Precalculus Through Applications* (Janson Publishing Co.), came from teachers at the North Carolina High School for Science and Mathematics.

Many innovative, effective efforts in mathematics preceding calculus are occurring at two-year colleges. To get around the daunting sequence of courses that stand between students with marginal mathematics skills and college majors such as biology or business that require calculus, some community colleges have developed various types of intensive courses. Seattle Central Community College has offered a one-semester 15-credit course that seeks to cover most of high school mathematics using an innovative curriculum and teaching style. Tech-Prep programs at two-year colleges are customizing mathematics for Associate's degree-level technical programs. A good number of these efforts involve integrated math/science/engineering courses.

7 Experiences in Learning Other Freshman-Level Mathematics

One of the interesting findings about the mathematics programs visited is that some of their students report their interest in becoming mathematics majors arose from freshman level mathematics courses other than calculus (or precalculus). At some institutions, such as Mt. Holyoke and Michigan, there has been a substantial faculty investment in developing freshman-level alternatives to calculus that are meant to interest students in taking more advanced mathematics courses while also being a meaningful first and last college mathematics course for other students. At other institutions, service courses not intended for students considering a major in mathematics are found to have motivated some students to become mathematics majors.

A statistics course for non-scientists (with a prerequisite of two or three years of high school mathematics) can excite students who enjoy working with practical and interesting data sets, whether they think of themselves as being mathematically inclined or not. A number of institutions have popular statistics courses that attract some students who are simultaneously taking calculus. Such students sense that there should be good careers in statistics that will make use of their facility with mathematics, a facility which they previously thought had little practical use except as utilized in the study of another discipline such as engineering. Many faculty in the effective programs are actively looking for potential majors in introductory statistics courses.

Some mathematics departments have tried to introduce special freshman-level courses for mathematics majors, to be taken by mathematics majors in addition to calculus. None of the mathematics programs visited has such courses. A more successful response to this need appears to lie in calculus reform efforts which include open-ended projects and mathematical modeling.

The special efforts at Mt. Holyoke and Michigan merit further discussion. Faculty at both institutions have been involved in major calculus reform. They also wanted to have interesting freshman level mathematics courses for liberal arts students not interested in calculus. They were interested in alternatives to calculus for potential mathematics majors. Mt. Holyoke faculty were also concerned that the long string of sequenced prerequisites discouraged many students, particularly female students, with an interest in mathematics from considering a major in mathematics. These long sequences meant that students do not really know much about the upper-division mathematics that is at the heart of the major when they declare their major.

Interestingly, both institutions independently developed two freshman seminars for non-majors with the same themes, one in number theory and one in geometry. The Michigan courses have a lighter dose of mathematics and more historical and cultural contexts, while the Mt. Holyoke courses focus more on explorations into mathematical structure. At both institutions, the reactions to these courses have been very favorable. The Mt. Holyoke freshman seminars attract some mathematics majors looking for some non-calculus mathematics. At Mt. Holyoke, students can continue into the sophomore-level Laboratory course (discussed in the next section) after one of the seminars and move on to: i) a minor with little or no calculus, or ii) a mathematics major program in which they take calculus later as upper-division students.

Instead of a freshman liberal arts seminar followed by a sophomore-level Laboratory course, Michigan has developed a two-semester alternative to the standard calculus sequence that covers some calculus but focuses primarily on getting students to think for themselves mathematically and gain confidence in mathematical reasoning (the students are mostly Honors students or at least have had some calculus in high school). The topics covered include combinatorics and finite differences/dynamical systems. The contents of this course was the topic of a conference entitled Discovery and Experimentation in the Freshman Mathematics Curriculum held in the summer of 1994. At the time of the site visit, Michigan faculty were planning a second-year sequel to this course. It is worth noting that several highly regarded young mathematics researchers at Michigan have been leaders in instructional innovations such as this course.

Another Mt. Holyoke option for non-mathematically oriented students is a pair of general education courses. The first, called Case Studies in Quantitative Reasoning, involves three case studies using mathematical and statistical methods in nonmathematical contexts. The three case studies in a recent year were "The Salem witchcraft trials: wealth and power in Salem village, 1681–1696," "SAT scores and predicting GPAs," and "Modeling population and resources." Each week, students attend one lecture, two

discussion sections and a 3-hour computer laboratory. Interestingly, most of the sections of this course are taught by non-mathematics faculty. Fifteen percent of the student body takes this course, several of whom become mathematics majors. The other, larger general education course, called Past and Presences, taught by a team of faculty from several departments, contains a 3-week segment on explorations in number theory (a subset of the material in the number theory seminar).

8 Experiences in Sophomore-Level Mathematics

After teaching calculus in a non-theoretical style, there is a need in sophomore courses to teach mathematics majors the theoretical skills required for upper-division mathematics courses. On the other hand, most non-majors in sophomore mathematics courses usually have little need to develop proof skills. A compromise seen in most programs visited, and in most mathematics programs generally, is to emphasize theory in linear algebra and emphasize applications in differential equations and probability/statistics.

An alternative is to have a Foundations of Mathematics course or similar course to develop abstract reasoning and proficiency with proofs. Spelman is typical of many liberal-arts colleges in requiring a foundations course that comes after four semesters of calculus with differential equations. The Foundations of Mathematics course at SUNY-Potsdam is used to develop pride as well as skill in proofs. The Potsdam foundations course seeks to develop thorough mastery at a relatively modest level of writing proofs, but it manages to include virtually all students in the positive experience.

Mt. Holyoke has a very successful sophomore alternative in the Foundation course called Laboratory in Mathematics, which has common features with the Michigan freshman alternative-to-calculus sequence. It covers a few topics in substantial depth. It stresses exploration of mathematics patterns with a heavy dose of writing (several 12-page papers) and extensive use of computers. Students said that in writing the papers, they were forced to understand the mathematical ideas for themselves—an experience that made seniors say this was the most valuable course they had in college. The course is very popular among students, many of whom said this course motivated them to become mathematics majors. Mt. Holyoke faculty report a measurable change in students initiative and mathematical reasoning in sequel upper-division courses since the Laboratory course was introduced.

Vanderbilt and SUNY-Stony Brook have been extremely successful in getting engineering and computer science students to choose a mathematics major with an applied bent on the basis of their positive experiences in required post-calculus courses in applied linear algebra, probability and discrete mathematics.

9 Attitudes Formed in Class about Future Mathematics Courses, the Mathematics Major, and Mathematics Careers

The mathematics programs visited were effective in developing positive attitudes about learning mathematics in introductory courses. They do little in their courses to try to inform students about subsequent mathematics courses or what the curriculum in a mathematics major is like. While they focus primarily on making the current course material engaging including mention of its uses, they are constantly asking questions which often cannot be answered without further coursework. In after-class and out-of-class conversations, faculty will encourage students to take further mathematics courses to realize their potential to excel in mathematics or simply take mathematics because they enjoy it.

There are many students who like mathematics but have preconceived notions about the lack of usefulness of mathematics that take time to overcome. Indeed, if students are good at mathematics in high school, many guidance counselors will tell them that they should consider a college major in engineering or computer science, since these are disciplines that depend heavily on mathematics and have good career prospects. Students who have gotten such advice in high school may enjoy their calculus course but think that there is no future in mathematics, while there is in their intended major.

Mathematics departments get a 'second chance' with such students because their majors frequently require one to three mathematics courses beyond first-year calculus—courses such as multivariable calculus, linear algebra, discrete mathematics, or (post-calculus) probability/statistics. At the time they take these mathematics courses, some students have started to realize that they do not enjoy engineering or computer science at the level of detail required by those majors. Rather they realize that they pre-

fer the mathematical problem-solving that underlies these disciplines. At universities and colleges with engineering schools, mathematics majors are often a minority in such courses, and the problem of striking the right balance between the converted—declared mathematics majors—and the non-converted—potential majors—can become particularly acute in these courses. The good instruction at the programs visited seemed to appeal effectively to both types of students.

There are a number of students who have another major but keep on taking one or two mathematics courses a semester because they like mathematics. At some institutions that attract above average numbers of mathematics majors, one finds a significant number of such students who eventually select mathematics as a second (double) major.

IV. Attracting Students to Study Mathematics: Out-of-Class Experiences

1 Experiences with Instructors During Office Visits

Students were frequently seen talking to faculty in their offices during the site visits. Faculty were generally available to students any time they were in their offices. It was common to see no official office hours posted. It was common for faculty to give out their telephone number and e-mail address to all students. Several students commented that permitting, and encouraging, such access made a very good impression with them.

During office visits, the instructor and student typically work together on a problem in a framework that seemed to put the two on an equal footing. Instructors felt that constructive, friendly experiences during office visits made students more comfortable in class about asking interesting, but risky, questions, such as 'why can't you approach the problem this way,' without fear of looking silly in front of classmates. Several students reported that office visits played a valuable role in their enjoyment of mathematics and made them more likely to continue studying it.

Office visits are important as a two-way street. At the same time that the student benefits from out of class discussions, the instructor is also benefiting. The instructor gets feedback from students about what concepts and exercises are causing difficulty. The instructor gets to know students as people and may find ways to assist students on matters not directly related to mathematics. When one reads the site visit reports in the Appendix of this study, one repeatedly finds references to the great personal interest that faculty showed for their students at these programs and the positive response this evoked from students. For example, much of the discussion at a faculty meeting at Spelman that visitors sat in on was devoted to concerns about students. Such personal connections between faculty and students are built largely in one-on-one out-of-class discussions, most commonly during office visits.

Office visits can be particularly beneficial for women and underrepresented groups. As noted earlier, these students are more likely to be insecure about their mathematical skills and more affected by the attitudes of instructors towards students. A friendly, caring instructor can make a big difference in interesting these students mathematics.

It was clear during the site visits that when instructors felt that they were "reaching" a student and knew that they were making a real difference in the student's interest and self-confidence about mathematics, they felt very good about themselves and this feeling carried over to their dealings with all students.

2 Advising About the Mathematics Major

At all the four-year colleges visited, all faculty share in the advising of mathematics majors. Often it was a collective advising of all majors by all the faculty. At the comprehensive and doctoral universities visited, a selected subset of concerned faculty did the advising. Some of the university mathematics departments had tried having all faculty do advising and found that some faculty did too poor a job to justify universal advising. When faculty talked about major advising it seemed clear that the advising sessions were used as opportunities to get to know the students better as people.

Students considering a major in mathematics will sometimes drop by the mathematics department office for information rather than approach a professor. During all site visits, office staff were friendly and respectful towards students. The University of Southern Mississippi (an institution not visited in this project) takes this policy almost to an extreme. For example, a member of the advisory committee reported that the department chair interrupted a discussion with the visitor to go help a student he saw waiting in the anteroom of the department office (while a receptionist was occupied on the telephone). A policy of going to great lengths to serve students outside of the classroom was credited at Southern Mississippi as leading to a large increase in mathematics majors as well as an increase in job satisfaction among faculty.

St. Olaf has an excellent method for bringing together the department's and the student's visions of what a mathematics major should be. The creation of their contract major, started in the late 1970s, has repeatedly been cited by St. Olaf mathematics faculty as the cornerstone for their highly successful mathematics major which for many years has graduated the highest percentage of mathematics majors of any college or university in the U.S. (around 12%). Miami University of Ohio has a modified contract system in which there are some prescribed courses and a student submits a plan for the major that is reviewed by a three-person faculty committee.

At St. Olaf, a prospective mathematics major and an advisor meet at one or more sessions to "negotiate a contract for the student's major. The student may want to take mostly computer science related courses with little abstract mathematics, such as abstract algebra or analysis, or vice versa. The advisor s job is to argue for a contract with adequate balance between application and theory.

Such dialogue forces faculty constantly to justify the value of the different components of the mathematics program and to be respectful and sensitive to the viewpoints of their students. In a sense, what is important at St. Olaf is not the contract, but the *contact*. Students are treated with respect by the very nature of the contract process and forced to play an active role in deciding what their mathematics major should be. Either side always has the option of saying the contract proposed by the other side is unacceptable–with the result of no major being developed for the student. In reality, it is extremely rare for a contract negotiation to fall through.

In this context, when a student agrees to take a theoretical or an applied course that she/he originally sought to avoid, the student will enter the course with a much better state of mind than if the course was a rigid requirement of the major. Likewise, faculty are cognizant that most courses have a goodly number of students who had to be talked into agreeing that this course might be valuable for them and the faculty's desire to vindicate such advice gives them extra motivation to reach these students in their teaching. It should be noted that most St. Olaf contracts tend to fall into a few general types, but it is important that the students discover these types for themselves.

For more information about advising, see "A Career Kit for Advisors," by Andrew Sterrett in the October 1994 issue of *FOCUS* and "Career Advising for Mathematics Majors,"

by David Lutzer in the October 1994 issue of *FOCUS*. The MAA Committee on Advising is developing materials to help faculty with advising. (For further information, contact the committee chair, David Lutzer of the College of William and Mary.)

3 Conversations Among Mathematics Majors

The programs visited had many mechanisms, some explicit and some implicit, to get mathematics majors to talk to one another. Cooperative learning, group projects and free-ranging discussion in class get mathematics students to know one another. Almost all departments visited had some sort of common room that undergraduates could use when faculty meetings, etc. were not occurring.

The most credible testimonials to the attractiveness of a mathematics major come from current mathematics majors. Upper- and lower-division students have many opportunities to mix and become friends in departmental extracurricular activities, and it is common for younger students to ask older students for academic advice. When a mathematics program is primarily serving a narrow group of students, e.g., prospective school teachers, the current majors will tell other students that mathematics is a good major if you like mathematics and want to teach. If the current majors enjoy their mathematical studies, see them as leading to a wide variety of career options and feel the faculty make learning mathematics very satisfying, then current majors will pass along to other students a very attractive image of the mathematics major.

The advice from upper-class majors to prospective majors is in writing at the University of Chicago. A committee of mathematics majors assembled by the College Dean produces a booklet helping students decide which mathematics courses to choose (similar booklets are produced for other disciplines). This is an important vote of confidence in students maturity and judgment by the University of Chicago administration.

Along with chance extracurricular meetings between current and prospective mathematics majors, some of the mathematics departments visited organize social /informational events for prospective majors to mix with current majors. It could be standing invitations to lower-division students to attend meetings of the mathematics club. It could be beginning-of-the-semester barbecues for all "friends of mathematics or recruitment events. Typically these events were organized and run jointly by a group of current students, assisted by one or two faculty. Planning such activities with fellow students and faculty is an excellent way for students get to know each other better.

Another way that upper-division majors were seen to interact with lower-division students was through tutoring at a departmental Mathematics Help Center. Even strong lower-division students will visit these centers occasionally. Working in these centers is very enjoyable for upper-division students who feel pride in assuming faculty-like roles. At some institutions visited, particularly at the University of New Hampshire, these centers were gathering places for current and future mathematics majors: there are quiet times when the tutors talk with each other and with friends who drop by to chat. At departments with active teacher preparation programs, such as New Hampshire s, this tutoring can be a valuable part of the pre-service teacher program.

There is a momentum that effective, high enrollment programs develop by having a critical mass of majors whose visibility in numbers makes other students look at mathematics as a natural major for them to consider if they have any disposition toward mathematics. Conversations with students at visited programs often revealed a sense that being a mathematics major was "in." (This contrasts with many experiences of site visitors on external reviews at other mathematics programs where students were a bit defensive when asked why they chose a mathematics major.) With a large number of majors, there is a high probability that among the majors there will be a group of "live wire" students eager to keep the mathematics club active and plan social events for the mathematics department.

4 Social and Informational Events

When this project first contacted the St. Olaf Mathematics Department for a possible site visit, the contact person was asked what was the key to the success of the St. Olaf mathematics program. The person thought for a while and then responded, "We love to eat!" This cryptic answer was followed shortly by the explanation, "Whenever we see a student in the halls, we think food!" The St. Olaf faculty are always finding excuses to organize pizza parties, ice cream socials and other casual get-togethers among faculty and their students, majors and non-majors.

As noted above, it is desirable for social and informational events to be organized and run jointly by students and faculty. At the programs visited, informational sessions for potential majors tended to have as much of a social component as possible. The technicalities of major requirements along with career information were presented in informational booklets. Any formal presentations were short, typically involving at most one faculty member along with a couple of students. Most of the time was devoted to answering questions followed by informal conversations over soda and snacks involving current majors, potential majors and a few faculty. At some effective programs, there are no recruitment information sessions. It is assumed that information about the major, e.g., course requirements, different tracks, and what one does with a mathematics major, has been addressed by faculty in office visits and in information booklets. Rather these schools focus on social events that are designed to allow potential majors to mix with current majors such as at a beginning-of-the-year barbecue. At institutions where many students are commuters, a luncheon barbecue may be more convenient.

Many institutions visited ran informational events about graduate study. Chicago has a series of talks by visiting mathematicians about graduate study in their institutions and by faculty from other quantitative departments about research and advanced training in their disciplines. Southern University has run MATHFests which bring in successful minority mathematical scientists from industry and academia and minority graduate students.

Social events for mathematics majors are discussed further in Section 6 of Chapter VI.

V. Attracting High School Students to Study Mathematics

The objective of high school outreach is to nurture interest in mathematics (irrespective of what institution a student attends), that is, to motivate students to take more mathematics wherever they go to college. As one member of the project advisory board put it: "We are all each other's recruiters." High school outreach for the sole purpose of recruiting prospective mathematics majors to one's institution is generally ineffective because students pick colleges for many reasons that usually have little to do with the quality of particular departments. Outreach should be a multi-faceted, sustained effort. Virtually all of the institutions visited in this project had outreach programs for students and/or teachers in their region.

The mathematics seen in high school is typically 300 to 2000 years old and can appear to be largely irrelevant to the modern world. Outreach efforts to schools at institutions visited sought to counter this impression and to convey mathematics as alive, interesting and highly relevant. To this end, many effective programs place heavy emphasis in outreach/recruitment materials on attractive mathematically oriented careers available to Bachelor's graduates. Lebanon Valley College puts a primary focus on actuarial careers and offers an actuarial science major along with the regular mathematics major. This is a sort of 'come on': most freshmen interested in mathematics declare the Actuarial Sciences major initially but end up in the regular Mathematics major.

Most programs visited with high enrollment mathematics majors have a correspondingly high number of incoming freshmen expressing an interest in mathematics. Many of these students were encouraged to come to the institutions to study mathematics by high school teachers who graduated from the institution, who had previous students who went to the institution and reported back their happiness with the mathematics program, or who had had pleasant interactions with mathematics faculty through outreach programs. That is, they have a lot of "satisfied customers."

University of New Hampshire and St. Olaf both have programs to bring one or two practicing K–12 teachers to their departments each year to be lecturers in precalculus and calculus courses as well as to help in the teacher preparation programs. The teachers return to their schools to be active salespeople for the institutions where they taught.

Further ideas for outreach to schools can be found in *A Source Book for College Mathematics Teaching*, by Alan Schoenfeld (MAA, 1989; pages 11–20).

1 High School Visits by Faculty

Several faculty said that an important aspect of high school visits is building contacts with mathematics teachers, for teachers' recommendations of one's institution and one's department to their students can have a major influence on students college choices. These contacts are a two-way street: mathematics teachers can encourage their students to consider going to one's institution, while faculty can recommend a good continuing education course at their institution, supply helpful supplemental materials, or help mentor a talented student.

Lebanon Valley College mathematics department has one of the most active mathematics recruitment programs in the country because the college admits most students directly into the major at the time of college admission. Faculty on school visits talk about mathematics generally but also encourage students to come to one of the monthly High School Visiting Days at the College.

For mathematics departments with a teacher preparation program, school visits are an integral part of the placement and observation of student teachers. Using these visits to talk to groups of high school mathematics students is usually welcomed by teachers. Even if one does not visit classes, one can leave informational materials about mathematics.

2 High School Class Visits by College Students and Alumni

The best person to speak about studying more mathematics is a current mathematics major who graduated from the school (or area) being visited. This person can share her/his interest in studying mathematics in terms with which high school students can identify. The mathematics major will be more knowledgeable in answering questions about student-oriented issues outside the classroom that are likely to arise (such as, what is a dorm life like at your institution?). For reasons mentioned in the previous subsection, it is useful to have a faculty member present also. Further, a student standing beside a professor provides a visual image of the friendly atmosphere among students and faculty at your institution.

Mathematics alumni who now are leading successful professional careers are live testimonials to the broad usefulness of a college major in mathematics. Particularly impressive examples include an actuary, a business executive in a job that seems to use little mathematics, or a mathematics graduate with an engineering or computing job title. Further, they can enrich any high school mathematics class with interesting examples of how they use mathematical thinking in their jobs.

Current majors and alumni are usually honored to be invited to participate in such visits and the experience strengthens their pride in their mathematics training. Faculty at several institutions visited mentioned that they encourage current majors to return to their high schools during college breaks and talk to their former high school teachers about their mathematics studies in college as well as to express their appreciation to the teachers for their (hopefully) enjoyable high school mathematical experiences.

If a department has good information about the location of alumni, the alumni can be used to meet with mathematically oriented applicants to your institution who live some distance from campus. General college recruitment typically draws networks of alumni. The Lebanon Valley Mathematics Department uses alumni who are actuaries to interview applicants who express an interest in mathematics. The names and addresses of all examinees who successfully pass each actuarial exam are published in an actuarial newsletter. The actuary professor in the Mathematics Department reads these lists in the newsletter to update addresses of actuarial graduates. He sends them a department newsletter from time to time and asks particular alumni for help in recruiting applicants who live near them. Successful actuaries have proven to be very effective recruiters. The professor noted that it is also useful to maintain contact with actuarial alumni to get help in obtaining summer actuarial internships for current mathematics majors (of course, such ties also promote alumni donations).

3   Educational Outreach Programs

As noted above, virtually all the institutions visited had various types of educational outreach programs for students and teachers in local schools. Many sponsored events at their institutions for high school students were some mix of a mathematics competition and a mathematical enrichment program. Organization of such events frequently drew heavily on the local mathematics club. In some cases, while the students were taking tests, their teachers were getting continuing education lecturers.

Several mathematics departments ran summer and Saturday enrichment programs for students. Others had programs for teachers on topics such as use of technology and implementing the NCTM *Standards*. Southern University has a variety of cooperative efforts with local schools, the largest being its Upward Bound program. University of Chicago faculty had outreach programs to school teachers and summer programs for talented high school students. St. Olaf has an NSF funded project for helping school teachers use the *Geometer's Sketchpad* software in their classes. Miami University of Ohio annually runs a school mathematics education conference. While these programs were typically self-supporting efforts with small budgets, organized by a couple of faculty or a student club as a mathematical public service for the schools, such activities do well for the institution, in terms of publicizing its mathematics program and developing potential applicants interested in mathematics.

A good source for further information about developing programs for minority students is the MAA's SUMMA Program. For information about programs for mathematically talented students, a resource is the Mathematicians and Educational Reform network, headquartered at University of Illinois at Chicago Circle Mathematics Department.

4   Recruitment Mailings

Recruitment materials at programs visited for prospective students were mostly personal—about the students, the faculty, the graduates, the learning atmosphere, and opportunities for faculty and students to interact. The major requirements were presented in a formal listing but also in the form of semester-by-semester course programs of typical majors with different goals, e.g., preparation for graduate school, pre-business, pre-actuarial, etc. Finally, some interesting mathematical exposition was sometimes included such as a copy of the MAA's new student publication *Math*

Horizons or an elementary expository article by a faculty member.

Mailings of a department newsletter to mathematics graduates who are now school teachers are very useful. The SUNY-Potsdam Mathematics Department, which trains many high school mathematics teachers, has used a newsletter effectively to stay in touch with alumni in secondary education and to encourage them to recommend Potsdam to their students. The newsletter contains information about what recent graduates are doing, recent departmental happenings, and the like. College alumni offices usually are very helpful in efforts to contact alumni and often will pay mailing costs to alumni of newsletters since these newsletters advance the broader college goal of staying in touch with alumni and building goodwill with alumni.

5   Campus Visits

A successful approach of several effective programs is to have high school students and their parents sit in on freshman mathematics classes (classes where there is likely to be a good amount of class participation) for an hour and then assemble to hear brief presentations from some faculty and current majors. Then visitors chat with majors and faculty over some food at tables. (This requires that the open house be during a class day.)

VI. Organization of the Mathematics Major and Supporting Activities

A program that is effective in attracting and graduating mathematics majors must be sensitive to local circumstances, e.g., whether the institution has a student body with a predominantly liberal arts outlook, or a professional, career-based outlook. Recall the precept of teaching the students one has, rather than the students one wishes one had.

1 Program Goals and Structure of the Major

The typical mathematics program visited had a middle-of-the-road, inclusive goal of a rigorous training in core mathematics, e.g., linear algebra, analysis, and algebra, supplemented with a choice of electives that range from pure mathematics to a variety of applied mathematical sciences topics. These electives reflect a flexibility in the major requirements, with a modest number of core courses required of all majors, and the existence of a number of different tracks within the mathematics major. This type of major is believed to be the norm at most mathematics departments today and seems well suited to the times and the current students. The programs visited took this flexibility farther in part because they had more majors who generated substantial enrollments in a diverse set of courses. On the other hand, some successful mathematics programs have much narrower objectives.

Middle-of-the-road programs can serve a variety of diverse student objectives: joining the workforce immediately upon graduation, continuing to graduate school in mathematics or applied mathematical sciences, or going to professional schools. Most of their upper-division mathematics courses are taught as a terminal course in that subject, with enrichment for those planning graduate study.

Some institutions, such as Miami of Ohio and University of New Hampshire, obtain sizable enrollments in their mathematics programs by combining a strong teacher preparation major with a traditional mathematics major. The two majors together can provide the critical mass needed for extensive upper-division offerings, for extracurricular student-run events and mathematics clubs, etc., and

for getting faculty involved in the undergraduate program. At noted earlier, Lebanon Valley College effectively combines an actuarial science major with a regular mathematics major. The former helps the department attract a large number of students to mathematics most of whom in time switch to the mathematics major.

At some popular mathematics programs with seemingly broad goals, faculty interests tend to focus students in narrower directions. Williams appears to be such an example, with a broad curriculum and many options in recruitment materials but with the majority of its graduates entering doctoral programs in mathematics.

Some effective mathematics programs have more narrow explicit and implicit goals. The University of Chicago is interested in preparing its students for doctoral study and makes this very clear to potential majors. Fifty percent of Chicago mathematics majors pursue graduate study in mathematics, and another 25% go to doctoral programs in related disciplines. At the other extreme are the SUNY-Stony Brook applied mathematics major and the Vanderbilt mathematics major which are designed to be terminal programs for students going to work or professional school upon graduation. The Chicago focus on pre-doctoral training is discussed extensively in Chapter VII.

The popular Vanderbilt mathematical sciences major (which is chosen by 10% of Vanderbilt's students) is pragmatically designed. It serves primarily engineering students who chose mathematics as a second (double) major and quantitatively oriented economics students intent on business careers (the institution has no undergraduate business program). The Vanderbilt major requires only about 30 semester credits including calculus (about 15 credits beyond the mathematics requirements in an engineering major) with a limited amount of theory. While some mathematicians may feel the Vanderbilt major is not rigorous enough, it provides an appealing package for training in mathematics for its students, the vast majority of whom would have taken only 6 to 15 credits of mathematics otherwise. As an alternative track in a mathematics program, such an option may be worth considering at any institution with professional schools in business and/or engineering.

17

The Mt. Holyoke mathematics major has an interesting organizational theme—bringing students in contact with advanced mathematics quickly. This involves freshman seminars, with abstract mathematics, that enable students initially to bypass the calculus sequence and move quickly to topics courses on advanced mathematics subjects with minimal prerequisites.

On the other hand, Potsdam resists the idea of a curricular theme. The chair at Potsdam once asked an MAA curriculum leader, "Why does the MAA spend so much time worrying about having the right syllabi and right curriculum. What does that have to do with training mathematicians?" All effective programs are foremost trying to train their students to reason mathematically.

A mathematics department that is trying to take stock of its undergraduate program is well advised to undertake a systemic program assessment. (More and more public institutions are being required by state legislatures to undertake such assessments.) For assistance, readers should see the CUPM report, *Assessment of Student Learning for Improving the Undergraduate Major in Mathematics,* in the June, 1995 issue of the MAA newsletter *FOCUS.*

2 Leadership and Faculty Participation

About half of the mathematics programs visited could point to major reworking of the undergraduate mathematics program sometime in the past two decades that launched them on their current successful course. Many other successful mathematics programs (with large enrollments) which have come to the attention of the project advisory committee have had such "turnarounds." These changes were typically led by a few faculty who managed to engage most of the rest of the faculty in these reforms.

Two-year and Four-Year Colleges   At 2-year and 4-year colleges, effective programs normally involve most of the (full-time) faculty and these faculty all usually participate in major decisions about the program as well as many lesser day-to-day decisions. At such institutions, many people over a period of 10 or 15 years have played important roles in the mathematics program. On the other hand, at some there are one or two people who have had a dominant role in putting in place and overseeing the mathematics program. Most mathematics programs visited fall in the former category while Lebanon Valley's program and the

well-known program at Potsdam, developed by Clarence Stephens, are examples of the latter.

Many successful 2-year and 4-year mathematics programs have effective chairs who are excellent at keeping the faculty happy, mentoring junior faculty, heading off personality conflicts among faculty, and generally serving like a department concierge. They create a departmental climate which facilitates the faculty efforts in effective programs. The chair's job is often distinct from being a curriculum innovator or leader in the instructional program. At 2-year colleges, the chair and senior faculty play an important role in overseeing how the many part-time faculty are integrated into the department. Austin Community College has been particularly effective in supervising and getting the best out of its part-time faculty.

Doctoral and Comprehensive Universities   The faculty at these institutions have a major commitment to research and graduate education. These institutions tend to have large service teaching. Lower-division courses are either taught with large lectures or in sections taught mostly by TAs. In this environment, it is fairly uncommon to find the majority of faculty engaged in innovative undergraduate instruction.

However there are a number of universities with a substantial number of faculty concerned about their undergraduate teaching. In doctoral institutions, the leadership of the department chair or a core group of faculty deeply engaged in the undergraduate program seems to be critical in creating the overall climate that values an excellent undergraduate program. The doctoral mathematics programs visited in this project—Chicago and Michigan—have such leaders who encourage, coordinate, and monitor faculty teaching efforts and work closely with students. These leaders champion the cause of the undergraduate program in departmental decision-making, e.g., to keep less effective teachers out of key undergraduate classes. Michigan had a substantial number of research faculty engaged in curriculum development and/or using instructional styles that require much more time than traditional instructional methods.

3 Content and Teaching of Courses in the Major

Standard courses such as linear algebra, introduction to analysis, and probability at the programs visited generally follow the standard syllabus used in most mathematics departments. The programs visited have more topics courses, in part because there are enough students to justify such courses, but also because faculty want to offer enrichment beyond the regular curriculum and are willing to run small topics courses for a couple of students as a voluntary extra course.

The teaching of upper-division mathematics courses in these programs tends to contain less lecturing. There is a greater effort to involve the students in the development of the material, most often through questioning of the students to lead them to develop new concepts, to frame correctly a definition or the statement of a theorem, and to outline the proof of a theorem.

The programs visited had a great variety in the teaching styles of the faculty and also variety in what topics get major emphasis in a course. What the faculty have in common is a sense of excitement about their teaching and an ability to engage their students. Everything else is quite individualistic. Common final exams in multi-section, lower-division courses were less common, because of the greater responsibility given to individual instructors.

Information about MAA recommendations for the content of mathematics courses can be found in the 1981 CUPM Report, *Recommendations for a General Mathematical Science Program* (reprinted in *Reshaping College Mathematics*, L. Steen, editor, MAA, 1989) and the 1991 CUPM Report, *The Undergraduate Major in the Mathematics Sciences.*

4   Major Requirements

The beginning core requirements of virtually all American mathematics majors consist of a three-semester calculus sequence (including multivariable calculus) and linear algebra. Differential equations is often also required, especially if the first linear algebra course occurs as a junior level course. At the upper-division level, there is much more variety now than there was 15 years ago. Most but not all mathematics majors require a semester of abstract algebra and a semester of advanced calculus/introductory analysis, but little beyond that is uniformly required. Additional required courses often are to be chosen from sets of courses, say, in analysis and in applied mathematics. While some mathematics programs prescribe almost all the courses in the major, the programs visited in this study generally had a limited number of required courses. Also they usually

had multiple tracks that gave students additional flexibility. At several institutions, some courses outside mathematics could be used as part of the upper-division major requirements. Several had a variety of dual or double major options. In general, the programs visited displayed an inclusive spirit that accepted a diverse set of programs of study in mathematics as valid study. Many required some special-type experience such as a senior seminar, independent reading, or a senior thesis.

5   Seminars, Independent Study, Internships

Most effective programs make a point of offering supplementary educational experiences for their majors. During the academic year, there are seminars, taught in the Oxford tutorial style; student colloquia, where students present their own research and give expository talks to other students; independent study; directed reading; internships; and assisting on faculty research projects. During the summer, there often were industrial internships and REU-type summer research programs (REU refers to the NSF program, Research Experiences for Undergraduates). Independent study in the form of a senior thesis was often required for an honors degree, but few programs required a senior thesis for all students. The programs visited often required all students to take a senior seminar course. The other supplementary experiences are generally offered as an option rather than a requirement. Several programs have semi-independent study experiences in some regular courses in the form of course projects.

The Michigan mathematics department is seeking to place every mathematics major either in a summer industrial internship or a summer research experience (with a stipend from REU funds). The Michigan research experience involves 'vertically integrated research teams' consisting of one or more senior faculty, some junior faculty, some graduate students and some undergraduates. Such teams are common in experimental sciences, but are relatively new to mathematics. These teams provide enhanced research environments to the junior faculty and graduate students as well as the undergraduates.

The women's colleges visited, Spelman and Mount Holyoke, put a particular emphasis on providing organized independent study experiences for their students, which often involved REU's at other institutions. For example, Spelman has arranged research opportunities at Bryn Mawr.

The students at the visited programs who undertake independent study are encouraged to present the results of their work at local mathematics club meetings and at sectional MAA meetings. Miami of Ohio runs a fall collegiate mathematics conference and a spring mathematical education conference which is heavily attended by Miami undergraduates, a number of whom present papers.

Summer research programs and independent study projects were cited by many mathematics students as playing a critical role in interesting them in advanced study in mathematics. On the other hand, Chicago has little organized independent study while producing a high percentage of majors who pursue doctoral study in mathematics. As noted earlier, it is common for these supplementary educational experiences to be undertaken voluntarily by faculty; they are not part of the faculty's official teaching load.

6 Mathematics Clubs

Most institutions visited had one or more active student mathematics organizations. These included departmental mathematics clubs, MAA student chapters, Pi Mu Epsilon chapters, teams preparing for the Putnam Competition and the Mathematics Modeling Competition. At the University of Chicago the student groups functioned without any formal faculty input, even the Putnam team. The mathematics club meetings had: presentations by students—both research talks about independent study projects and expository talks about interesting papers in expository journals such as *Mathematics Magazine*; presentations by local faculty and visiting faculty; talks about careers and industrial applications by alumni and speakers from industry. Student talks at clubs seemed to instill considerable pride and mathematical self-confidence in mathematics club members.

The mathematics clubs usually organized several social events each year, at least one targeted at prospective mathematics majors. The mathematics clubs were also a source of organized student input to the departments.

7 Student-Faculty Interactions

Social events are very important to help develop comfortable relations between students and faculty. Comments by students during the Lebanon Valley site visit illustrate the importance of 'comfort': students said they liked their small mathematics department where in time students would have most faculty for two or even three courses, because they said they felt more comfortable in a class when they knew the professor from a previous course. Out-of-class social contacts can be equally effective in developing such comfort.

It should be noted that at many of the programs visited student-faculty contacts often continue, by letter, e-mail and return visits, long after students graduate. For example, during the visit to Spelman, an alumna who was completing her MS elsewhere was visiting the department for advice as she prepared to go into the job market.

Students who know their instructors well are more likely to give constructive feedback to faculty about their instruction and the overall mathematics curriculum. When faculty are constantly experimenting with their courses, such feedback is essential. Also at programs that had a dramatic re-working of the major, student feedback was essential when faculty were trying to determine which new approaches were working.

Most of the programs visited have mechanisms for extensive student input in departmental decision-making. The input comes through either a student advisory board, whose members are elected by students, or the officers of a student mathematics organization. Most of the liberal arts colleges visited have students participate in interviewing faculty candidates. At such institutions, student input in evaluating teaching is very extensive; some require input from alumni in tenure decisions.

At institutions located in college towns where faculty typically live close to campus, faculty frequently invite students over to their homes for informal get-togethers and larger parties. The very nature of a home is highly conducive to easy socializing and allowing people to get to know each other without invisible tags saying 'I am a student (or a professor).' Monthly or weekly faculty-student dinners seem to be more common at small colleges visited.

The facilities and location of a mathematics department are also important. Several programs visited give considerable importance to having a student study area around the mathematics department offices, preferably with a small department library with reference materials for undergraduate courses. Such space can be a gathering place for mathematics majors to study and socialize. It builds a physical sense of belonging to the mathematics department. Since faculty tend to drop by this study area frequently, it also facilitates student-faculty contacts. The Michigan Mathematics Program recently established a special area for mathematics majors with computing facilities and an academic advising office with a full-time staff member. At Williams College, most mathematics majors use the mathematics library extensively for studying, and the mathematics department sees this as a major factor in promoting a high level of student-faculty interaction.

Most of the programs visited have mathematics classes taught in rooms in the same building—many on the same hallway—as faculty offices. These hallways are often filled with posters and display cases of a mathematical nature. This heightens contact between faculty and students before and after classes. It seems likely to increase visits to offices, since going to a faculty office would seem less forbidding when one passes these faculty offices regularly going to and from class.

8 Relations with Other Departments

As well as serving students well, the programs visited generally received praise for their instruction from faculty in science departments and other departments whose majors are served by the mathematics courses. Since many mathematics majors start in other programs, effective instruction for students in other majors is almost synonymous with effective instruction for mathematics majors. Moreover, such effective instruction carries over to effective interactions with faculty in other departments. These interactions involve frequent consultation about their teaching in service courses and jointly developed or taught courses. Clearly, advisors in these departments would be more likely to encourage their students to take additional mathematics courses. The great care that most of the departments visited put in their mathematics placement exams resulted in the exam scores being used for placement by some science departments—such accurate assessments were greatly appreciated.

At Mt. Holyoke there was some dissatisfaction with the content of the Calculus in Context course. Yet on many other fronts, the mathematics faculty was cooperating effectively with other departments. A disagreement on curricular priorities in calculus that might poison relations at many institutions was just a problem among friends at this institution.

9 External Funding for Instructional Activities

Most of the mathematics departments that were visited had received external funding from federal and private sources to help support technological and instructional innovation. Such external support has generated individual self-esteem and departmental respect for mathematicians at liberal arts colleges. However, the mathematics programs at Chicago and Lebanon Valley College (as well as the well-known SUNY-Potsdam program) have received no external assistance for instructional activities. These institutions have done the basics well as opposed to developing or adopting new curricula.

Unlike in research, there is little external support for excellent instruction. However, there is large internal support. For example, the St. Olaf Mathematics Department almost doubled in size as its contract major program took off. The Michigan Mathematics Department received a ten-year, $1,000,000-a-year commitment from its Dean to permit more sections of 'new wave calculus to be taught with faculty in small classes. The past chair of the Michigan Mathematics Department, Don Lewis, made the statement that research mathematicians have their priorities for support mixed up: "the two-ninths summer research support is icing on the cake; the cake is calculus."

10 Informational materials

Most effective programs have several publications with information about the mathematics major. These publications generally divide into two types. One is a sort of handbook about the major, including all requirements, seminars and colloquia, clubs and honor societies, summer programs, extra-curricular programs, computer labs, whom to see for help about various matters, etc. The second is a recruitment brochure which is a 'case statement' for majoring in mathematics at their institution. The latter makes the case:(i) for mathematics in general; (ii) for the satisfying experience of being a mathematics major at the institution; and (iii) for the rewarding careers and advanced study opportunities of recent graduates.

The University of Chicago's undergraduate handbook is written by the members of the undergraduate mathematics club. The handbook includes student assessments of the value, difficulty, intended audience, etc., of all mathematics courses. Some effective programs have a brochure just about careers, often profiling recent graduates and their careers. The MAA, the AMS and the Conference Board of Mathematical Sciences have produced glossy booklets profiling mathematical scientists which can be used for this purpose. The MAA Student Chapters program has a Mathematician of the Month profile. Note also the career articles mentioned earlier: "A Career Kit for Advisors," by Andy Sterrett in the October 1994 issue of *FOCUS* and "Career Advising for Mathematics Majors," by David Lutzer in the October 1994 issue of *FOCUS*.

Expanded (longer and larger print) versions of such profiles can be placed in a display case (and rotated from time to time). The same is true for display cases with information about recent happenings in mathematics—newspaper clippings of articles involving mathematics, expository displays of mathematical concepts and theorems. Another possibility is the cover pages of student papers presented at a sectional MAA meeting. It is also attractive to have mathematical posters on walls, such as the ones *Mathematica* has produced. Some mathematics departments have painted mathematical symbols on corridor walls around the department.

VII. Effective Programs for Preparing Students for Advanced Study

1 Two Types of Programs that are Effective in Preparing Students for Advanced Study.

Programs that are effective in preparing students for advanced study in mathematics are of two general types. The first type is programs with (relatively) large numbers of students which engage their students in learning mathematics as preparation for a wide variety of possible careers and advanced education. The large size of the programs and the attractiveness of the learning experience results in a number of students pursuing a PhD in mathematics who had not previously planned on advanced study in mathematics. Spelman and St. Olaf are examples of such programs.

The second type is programs which make preparation for doctoral study in mathematics their primary goal. The most successful of such programs have 50% of their majors enter doctoral programs in mathematics. These successful programs are at highly selective private institutions. Examples are Chicago and Williams.

Both types of successful programs are characterized by the attributes of overall effective programs mentioned in the previous chapter. The faculty teach very well, the students are active learners in class and have substantial contact with each other and with faculty outside of class. The class sizes are normally small to moderate. Multiple sections of post-calculus courses are scheduled if the enrollment is over 30. The faculty are not only effective at motivating students to share their love of mathematics but also are generally viewed as positive role models by students—the students would be very happy to grow up to be the type of people their instructors are. It may be just a coincidence, but students at these successful programs are treated as colleagues by the faculty—through invitations to faculty teas, planning social events together, working together in outreach programs to local schools, etc.

A common theme at such programs is the participation of most, or all, mathematics majors in some type of research opportunity—REUs, senior seminars, senior theses, and the like. There are also opportunities for students to present their research results—at mathematics club meetings, in journal papers, and at professional conferences. At colleges it appears more natural for faculty to become engaged in sponsoring undergraduate research since there are no doctoral students to work with. The one doctoral mathematics department that this project learned about with a high percentage of graduates going on to doctoral study in mathematics, Chicago, has several special factors which make it atypical of doctoral departments.

Chicago has a very large mathematics faculty, about 50, for a small university. This size allows faculty who do not like, or show little facility at, undergraduate teaching to be assigned to graduate courses. Further, the large faculty size, along with graduate students who are prepared for undergraduate teaching through a superb training program, permits very small course sizes. (The very extensive training program for graduate teaching assistants is a natural way that a doctoral institution can maximize its impact on good undergraduate teaching; see the Chicago site visit report for more information about this training program.) Also there is a university-wide tradition at Chicago of preparing undergraduates for doctoral study—63% of Chicago graduates go on to masters/doctoral programs (this number excludes professional programs such as law, business and medicine) that is unparalleled in this country. By comparison highly selective institutions, such as Yale, Princeton, Swarthmore, Amherst, have only 15% of their graduates going to graduate study, based on the 1993 US News College Guide.

While Chicago has half of its mathematics majors go on to doctoral study in mathematics, a higher fraction than any other U.S. university, it also graduates one of the highest percentages of Bachelor's degrees in mathematics of any U.S. university, according to Project Kaleidoscope data. The Chicago mathematics program is not an elitist program. Rather it is able to attract and motivate students who would seem to be poor candidates for graduate study in mathematics. For example, one recent Chicago mathematics graduate was not oriented towards mathematics but did take an AP calculus course in high school which she flunked. In summer school she took a calculus course at a local university which she liked better and did well in, but on the mathematics placement exam taken on entrance to Chicago, she did not do well enough to place out of the first semester of calculus and ended up taking beginning calculus a third time. While not planning on a mathematics major, the Chicago calculus course was so enjoyable and the mathematics advisor she spoke to was so encouraging and friendly that she kept taking more mathematics and chose it as a major. Once she was caught up in the demanding but supportive atmosphere of the Chicago mathematics department, she did B level work in abstract algebra and related challenging courses. Upon graduation she went to a Group I mathematics graduate program.

The mathematics program at Williams College recently has had over 60% of its mathematics majors pursuing graduate work in mathematics, although only about 10% of all Williams graduates undertake advanced study (excluding professional schools). Williams also has an above average number of students majoring in mathematics. Williams does not have a tradition of many graduates going to graduate school in mathematics or of many mathematics majors generally. The current success reflects efforts to revitalize the Williams mathematics program in the 1980's, led by Frank Morgan. The hallmarks of the new effort were undergraduate research experiences and a large amount of student-faculty out-of-class interaction, including two dinners every week where mathematics faculty and students talk mathematics.

As noted at the outset of this chapter, some of the mathematics programs with large total number of majors are sending a significant number of students to graduate study in mathematics. Some are at institutions with a tradition of predoctoral training. For example, Mt. Holyoke College has produced the most female PhD's in science in the 20th century. Likewise, Spelman College (which was founded by the Rockefeller family around the same time they founded the University of Chicago) has a tradition of preparing black women for advanced study. On the other hand, St. Olaf College has no such tradition and ranks as being only selective (not highly selective) in most college guides. However, in recent years St. Olaf has produced the fifth most mathematics PhD's among U.S. liberal arts colleges.

2 Creating an Environment Conducive to Advanced Study

Programs that are effective in preparing students for advanced study in mathematics help create an environment similar to the one students will find in doctoral programs. While a good curriculum that lays the mathematical foundation for advanced study is very important, the extracurricular aspects of a student's mathematics education appear to be equally critical. A conducive environment includes faculty who are dedicated to the values of mathematical research and who share these values with students. Even more, effective programs foster extensive mathematical discussions among students as well as between students and faculty. Virtually all PhDs would acknowledge that their study groups and mathematical discussions with fellow students, along with discussions with faculty, were the heart of the learning process in graduate school.

One of the hallmarks of some of the top mathematics PhD programs is a daily tea at which faculty and graduate students gather for about an hour and talk. At the Princeton mathematics program, the role of tea time discussions is taken very seriously. Princeton has virtually no standard graduate mathematics courses, just research seminars (doctoral students are expected to have taken beginning graduate courses as advanced undergraduates or to learn the material on their own). In the absence of grades in regular graduate courses to assess students, faculty have measured students progress by keeping track of their attendance at daily tea (there are also comprehensive exams). Students seen regularly engaged in mathematical conversations are judged to be making satisfactory academic progress, because experience has shown that such participation in the intellectual life of the department is highly correlated with success in moving towards the writing of a dissertation. Once a week, Chicago has a tea for undergraduates together with graduate students and faculty.

A department does not need a daily tea hour, but there is great benefit from the environment created by such a gathering. Mathematical discussions among students can readily flourish when there is a critical mass of mathematically talented students. For 4-year colleges (without graduate students), a critical mass of such students is likely to exist if the institution is highly selective. But most highly selective colleges and universities do not have the number of prospective doctoral students among their mathematics majors that institutions like Chicago do. The goals and mission of these other institutions have different overall atmosphere. Yet, some institutions that are less selective produce more than the expected number of mathematics doctoral students by working consciously to create this sort of environment. The level of the mathematical discussion may not be near the frontiers of current research, but the fact that substantial discussion among faculty and students is occurring increases the odds for going on graduate school in mathematics and earning a doctorate.

A conducive environment also means that faculty should motivate and help students learn in classes—using the carrot—rather than criticize and embarrass students who do not perform up to expected standards—using the stick. Demanding sophomore honors mathematics courses taken by talented incoming freshman at highly selective institutions are famous for traumatizing many of their students so much that most never take any more mathematics courses as undergraduates, even though these same courses may start some students on a path leading to successful careers as research mathematicians.

VIII. Effective Programs for Preserve Preparation of School Mathematics Teachers

College and university mathematics programs are in a symbiotic relationship with the NCTM *Standards* movement. Success in this movement should produce students far more able to think mathematically and better prepared to do college mathematics. However, such success is heavily dependent on preparing a new generation of teachers who are comfortable with the less structured, more inquiry based style of the *Standards* curriculum and instruction. New teachers must have the commitment to bring about the adoption of a *Standards*-based curriculum in the face of resistance to change among other teachers, students, and many administrators.

It is disappointing to note that the support that the NCTM *Standards* have received from most college and university mathematicians has not been matched by any significant change in the curriculum or teaching for prospective mathematics teachers. Except for calculus reform and some efforts being made to revise courses for prospective teachers, courses and curriculum remain pretty much the same. While students in preservice teaching tracks often are half or more of the mathematics majors in a good many mathematics programs, these programs normally reflect the training of the faculty rather than the needs of these students.

For this project, the advisory committee looked for programs where prospective teachers got more than the traditional mathematics major program whose course syllabi typically are designed as a warm-up for graduate study in the mathematical sciences. Also the teaching approaches and learning experiences of the preservice teachers should reflect the more exploratory learning styles encouraged in the NCTM *Standards*. Unfortunately, the advisory committee found that many institutions still need to review and revise their offerings for preservice teachers to meet these goals.

For preservice secondary teachers of school mathematics, the mathematics community, through the calculus reform movement, has given the *Standards* movement important support. In reformed courses, preservice teachers are engaged in the type of teaching and learning the *Standards* envision. The calculus reform movement's emphasis on multiple approaches (algebraic, graphic, numerical, and applications) and on various instructional styles has put pressure on high schools to make similar changes in their mathematics programs.

This project included site visits to two institutions, Miami University in Oxford, Ohio, and the University of New Hampshire, where there are strong teacher preparation programs. However at these schools, the core mathematics courses are taught largely using traditional syllabi. St. Olaf College has an active teacher preparation option in the mathematics major but there is only one faculty member associated with that option. This chapter includes, in addition to these visited sites, references to the mathematics program at Illinois State University where most mathematics courses are designed with prospective teachers in mind.

There are four components that should be included within the mathematics program for good teacher preparation. First, the curriculum and course syllabi should incorporate the needs of preservice teachers. Second, the instructional style in the department should serve as a model for good teaching practices. Third, the mathematics methods courses should take a broad view of contemporary mathematics education issues. Fourth, there should be out-of-class activities and interactions with mathematical education faculty that enhance the in-class education and foster an intellectually active atmosphere of inquiry about new educational trends.

The four components are addressed below. First we note that in many cases, these can accomplished within the mathematics department. In others, some assistance is needed from the School of Education. Whenever the latter is the case, there needs to be excellent working relations between the units. When the two do not get along, it sends a very negative message to preservice teachers.

1 Curriculum and Course Syllabi

The Mathematical Association of America's Committee on the Mathematical Education of Teachers (COMET) worked closely with the National Council of Teachers of Mathematics in drafting curriculum guidelines for teacher preparation for the NCTM *Standards*; see the 1991 COMET report, *A Call for Change: Recommendations for the Mathematical Preparation of Teachers of Mathematics.*

As noted above, the multiple approaches to learning in reformed calculus courses are helpful to preservice secondary school teachers. Calculus is of critical importance because the college preparatory mathematics curriculum in high schools has readiness for calculus as a major goal. Good preparation for calculus will probably continue to be the sina qua non for high school mathematics for some time. If calculus is stressing conceptual understanding and visual and numerical approaches along with algebraic skills, and if calculus courses have open-ended problems, more applications and extensive use of technology, then these preservice students are very likely to become comfortable with such approaches. Further, they will want their students to become comfortable with them in high school in preparation for calculus.

Miami of Ohio developed a creative way to bridge the traditional syllabus in abstract algebra with the needs of preservice teachers. They have offered a one-credit associated seminar course that connects topics in algebra with school mathematics (although this course is not offered every year).

At the University of New Hampshire, a large mathematics education summer program in which many mathematics faculty teach has sensitized the mathematics faculty to teaching styles that are appropriate for in-service teachers and this carries over to academic-year teaching. For example, one of the non-mathematics education faculty members recently taught the abstract algebra course by briefly touring the "algebra zoo of different algebraic structures and then settling down to focus the course in a single algebraic structure that was developed in the class slowly through a student-based discovery fashion. New Hampshire also has a research oriented, senior seminar required of all preservice teachers, in the spirit of the senior seminars found in regular mathematics majors.

At Illinois State, coursework in topics such as abstract algebra is structured to support both graduate school preparation in mathematics and teacher preparation. It develops modern algebra from rings to integral domains to fields to groups instead of discussing groups first. For individuals taking only one semester, they get a chance to see the development of the number systems, rings of polynomials and matrices, and deal with the classical problems that relate to school curricula. For those students continuing to the second semester which focuses on groups and additional field theory, the initial experience with familiar structures such as $\mathbb{Z}, \mathbb{R}, \mathbb{C}, \mathbb{R}[x]$, etc., gives them more maturity when they study groups.

Illinois State's introductory linear algebra course spends a greater amount of time on systems and geometric transformations than the usual introductory course to provide a strong background for teachers. The department offers a second course to complete the full story through the spectral decomposition theorem for regular mathematics students.

Both New Hampshire and Illinois State require more upper-division courses of preservice teachers than of regular mathematics majors. Their majors require core areas such as advanced calculus and abstract algebra and topics particularly relevant to school mathematics such as discrete mathematics, geometry, probability and statistics. For example, all Illinois State preservice students typically take, in addition to the preceding courses, a second semester of prob/stat or algebra and an applied mathematics course. Of course, they also have a number of methods courses and education courses. These heavy major requirements reflect a professional view of teaching, in the spirit of the professionalism inherent in the heavy coursework of an engineering major.

2 Instructional Style

It is obvious that how preservice students are taught and how they learn as undergraduates are going to be major influences on how they teach and how they expect their students to learn. The alternate modes of instruction in calculus reform involving group learning, extensive writing, and use of technology mirror instructional goals of the NCTM *Standards*. Students need experience in being active learners before they can enable others to be.

Like many issues studied in this project, innovative instruction that engages students is valuable for all types of students and all types of career goals. In virtually all the institutions visited in this project, much of the instruction engaged the students in various ways to become active learners .

3 Methods Courses

The methods courses for teaching mathematics at most of the institutions visited covered the mathematical roots and extensions of school mathematics as well as instructional approaches for teaching school mathematics. Such methods courses also covered technological applications, problem-solving, and curriculum/instructional materials development. All gave substantial coverage to the NCTM *Standards*.

It is highly desirable to have such courses taught by faculty with extensive experience in mathematics education. These are professional courses that should be taught by a professional. Faculty teaching these courses were usually engaged in the supervision of student teachers in the schools so that their classroom instruction is integrated with the students teaching experiences.

4 Supporting Out-of-Class Activities

New Hampshire and St. Olaf have mechanisms for bringing practicing teachers to their campuses to teach—and interact with students who are preservice teachers. Miami of Ohio runs an annual mathematics education conference and preservice teachers are encouraged to present papers on independent research projects in mathematics education. New Hampshire and Miami of Ohio have regular mathematics education seminars. Such seminars tend to draw a number of non-education mathematics faculty, since many of the topics discussed have relevance to collegiate instruction. Prospective teachers in their senior year often have the maturity to attend such seminars. Generally a lively faculty development program is important in mathematics education just as it is in core mathematics.

There are a number of activities that are not central to preservice teachers training but which enhance their college experiences. Most of the departments visited have outreach programs for the schools in which preservice students are heavily engaged: mathematics competitions, MathFests (a combination of competitions and enrichment programs), and Saturday and summer enrichment programs for students and teachers. Many mathematics departments encourage preservice teachers to be tutors in the department's help center. This can have the effect of turning the help center into a gathering spot for preservice teachers.

Finally, when there is a School of Education at one's institution it is essential for there to be good working relationships between the mathematics and mathematics education faculty so that programs for preservice teachers are coherent. Collaboration between these groups of faculty has mutual benefits and sends a positive signal to preservice teachers.

IX. Effective Programs for Underrepresented Groups in Mathematics

Mathematics programs that were effective in attracting and graduating members of underrepresented groups had the same general attributes as those programs that were effective in attracting and graduating substantial numbers of students generally. This illustrates the old adage that a rising tide raises all ships. However, studies by Sheila Tobias (*Revitalizing Undergraduate Science: Why Some Things Work and Most Don't*, report of the Research Corporation, 1990) have documented that effective instructional programs in the physical sciences and mathematics disproportionately benefit members of underrepresented groups. This case studies project found much support for that thesis in the mathematical sciences where women are concerned but little support where underrepresented minorities are concerned.

The project's set of site visits included two women s colleges—Mt. Holyoke College and Spelman College—and two historically Black colleges/universities (HBCU)—Spelman College and Southern University. The effective instructional practices and interactions with students found there were the same as those found at the other non-research institutions visited. The coed, non-HBCU institutions visited typically had nothing in the way of special programs for women or underrepresented minorities; the one exception was Michigan.

The critical role of effective teaching in S.M.E. [science, mathematics and engineering] retention efforts is documented in a recent Sloan Foundation-funded study of why students leave or remain in S.M.E. majors. It reaches the following conclusion: "switchers and non-switchers [out of S.M.E.] were almost unanimous in their view that no set of problems in S.M.E. majors was more in need of urgent, radical improvement than faculty pedagogy. All related matters, including the curriculum revision, were deemed secondary to this need. ... Though faculty sometimes like to begin a program of reform with discussions about curriculum structure and content, this is unlikely to improve retention unless it is part of a parallel, and iterative, discussion of how best to present these materials so as to secure maximum student comprehension, ... *Talking about Leaving*, Report to the Sloan Foundation, by E. Seymour and N.

Hewitt, April, 1994. This study found women and underrepresented minorities particularly affected by these non-curricular issues.

Implicit in the previous discussion is that minority efforts should be grounded in the disciplines rather than in some generic university-wide support center. The critical role of instruction requires academically based support programs.

The best known effort to enhance an underrepresented group in mathematics is Uri Treisman's Professional Development Program (PDP) at Berkeley which involved supplementary problem-solving workshops for calculus using a cooperative learning format. The workshop goals were to help students not just to pass calculus courses but rather to excel in them—earn A's or B's. Their problems were challenging and less cut-and-dried—many would now be classified as calculus reform types of problems. This program greatly increased participants' performances in calculus and their completion rates in S.M.E. majors at Berkeley.

The PDP workshops have been tried at other institutions with generally successful results. For example, CalPoly-Pomona has documented vastly improved completion rates for Black and Latino students who participated in the local PDP program. For an independent assessment of the PDP model, see Marty Bonsangue, "An efficacy study of the calculus workshop model," *Research in Collegiate Mathematics Education I*, CBMS Issues in Mathematics Education 4, pp. 117–138, 1994.

Yvonne Powell's "Students learn mathematics when they work collaboratively, *Mathematics in College*, p. 68–76, 1993 (*Mathematics in College* is an annual publication of the City University of New York) is an article that summarizes several studies, including the Workshop model, of the effect of cooperative learning on underrepresented groups. Again, the results are all positive.

A major survey of existing research (up to about 10 years ago) concluded that mixing ethnic groups in a school can be detrimental to minority groups, but that if certain cooperative learning strategies are used, the effect can be reversed. See David Johnson, Roger Johnson, and Geoffrey Maruyams, "Interdependence and interpersonal attraction among heterogeneous and homogeneous individuals: A theoretical formulation and a meta-analysis of the research, *Review of Educational Research* 53 (1), pp. 5–54, 1983.

1 Women

All the mathematics programs visited in this survey appeared to have above average percentages of women among mathematics majors compared to peer institutions, e.g., comparing Chicago to Yale or Swarthmore. A majority of St. Olaf mathematics majors are women. The very successful mathematics program at SUNY-Potsdam also has more women than men. In student discussions at the programs visited, female students did not mention incidents of biased behavior by faculty against women in instruction or out-of-class interactions, although a few women mentioned occasional sexist attitudes on the part of fellow male students. Rather, women felt they were seen by faculty as equals and participated equally in mathematics classes.

NSF studies (e.g., see the 1992 NSF report, *Indicators of Science and Mathematics Education*) have estimated the percentage of Bachelors degrees in mathematics earned by women at over 40% in the past decade. By this measure, the mathematics profession stands above the physical sciences and engineering. However, in mathematics departments where few, if any, students obtain preservice teacher training, there is substantial anecdotal evidence that the percentage of female mathematics majors is lower. At highly selective institutions, it appears to be 25% or less.

Interviews with female students in the site visits did not reveal any differences in the attitudes or expectations of women versus men, e.g., such as women deferring to men to play leadership roles in group projects, or one sex preferring professional (MS-level) training over doctoral training.

There is one difference in curriculum worth mention. The Mt. Holyoke mathematics faculty felt that the long sequence of prerequisites preceding courses on contemporary mathematics topics, such as Lie groups, was a liability in attracting women to major in mathematics. Consequently, they found ways to make such mathematics more easily accessible through shortened alternatives for reaching upper-division mathematics courses and specially designed courses in advanced topics with few prerequisites; see the Mt. Holyoke site visit report for details. However, based on the commonalty of effective techniques for women and men mentioned above, there is reason to believe that Mt. Holyoke's approach might attract more men to major in mathematics also.

2 Underrepresented Minorities

Unfortunately, the four-year mathematics programs visited in this project had negligible numbers of Black and Latino mathematics majors, except of course for Southern University and Spelman College, which are historically Black institutions. Blacks receive almost 5% of Bachelor degrees in mathematics and Latinos about 2 1/2%, according to the 1990 NRC report, *A Challenge of Numbers.* However, a substantial fraction of Black mathematics Bachelors come from HBCUs, so that at non-HBCU institutions, the percentage of mathematics BS's going to all underrepresented minorities is probably only a bit over 5%. These groups represent about 20% of the U.S. citizenry. There is much work to be done at non-HBCUs to achieve greater participation of underrepresented minorities. The programs that drew large numbers of other types of students apparently need to do something different and special for attracting and retaining this group in mathematics. A good source for further information about developing programs for minority students is the MAA's SUMMA Program.

The two HBCUs visited in this project had a very nurturing environment in their mathematics programs that arose from faculty who cared very much about their students. This atmosphere is typical of HBCUs whose historical mission was to develop their students academic and personal skills to succeed in a world where the deck was stacked severely against them. The heavy teaching loads at HBCUs help keep class sizes small.

The Southern site visit report noted the strong interest in the applied mathematics track in the mathematics major. The report mentioned that Southern mathematics majors were very career conscious: they wanted to be prepared for mathematically based professional careers upon graduation. The interest in professional careers is consistent with NSF data that show, for example, that freshman minority students are 10 times more likely to say they want to become engineers than scientists.

On the other hand, many Caucasian students these days are also very career conscious. Most of the large enrollment mathematics programs visited in this project had majors that were intended to offer good preparation for professional careers and had extensive career information. Several used this career information to interest students in the major and once in the major motivated students to have broader mathematics interests. This approach should work well for minority students.

A critical issue for many mathematics departments is

that their institutions would seem to need a critical mass of minority students with science/mathematics/engineering (S.M.E.) interests. Unless an institution has a history of solid enrollments of minorities in the physical sciences and mathematics, the presence of an engineering program appears to be a major contributor to the creation of a critical mass of minority students in freshman calculus. This seems to reflect the professional focus, mentioned above, of many minority students with good quantitative skills. In turn, the engineering community has obtained much corporate, government and foundation funding (particularly Sloan Foundation support) to recruit minorities into engineering and to help them succeed. Thus, mathematics departments can benefit from mathematics support programs designed for minorities in engineering, but which overall help create a critical mass of quantitatively oriented minority students.

However a critical mass of minority students is not enough. Michigan had a special program to help minority students, in the spirit of Treisman's successful PDP program at Berkeley. Like the Berkeley program, this effort was aimed more at helping students with calculus as a tool needed in engineering or science majors. There is substantial documentation of the effectiveness of such efforts generally, but little evidence of an impact on the number of minority students majoring in mathematics. On the other hand, the professionally oriented Applied Math program at SUNY-Stony Brook has seen its number of minority majors go from 2% to around 15% in the past 8 years with the introduction of entry support programs for minorities and Treisman-type calculus workshops, which were created to increase the number of minority engineering students.

X. Concluding Remarks

This report is a companion to general recommendations by the MAA and other groups on enhancing the undergraduate experience in mathematics by providing mathematics faculty with general attitudes and strategies, as well as particular activities, that promote effective instruction and learning in undergraduate mathematics programs. The efforts at institutions mentioned in the report are offered as examples to stimulate readers own efforts.

Attitudes towards students and interactions with students, inside and outside the classroom, seemed to be the primary factor in effective programs. The actual curriculum seemed to play a secondary role. As in the calculus reform movement, one may start collegiate mathematics educational reform with an eye on curriculum but one soon comes to realize that how students are learning is in many ways more important than exactly what they are learning.

There appear to be four major components of efforts to reform mathematics instruction.

Assessing the goals of the current program and aligning them with the needs of the students. As mentioned at the end of section VI.1, a starting place for program assessment is the CUPM report, *Assessment of Student Learning for Improving the Undergraduate Major in Mathematics,* in the June 1995 issue of the MAA newsletter *FOCUS.* Also see the commentary on challenges in program assessment in the November 1995 *FOCUS* column of the MAA Education Council. In many ways, the challenge in program assessment is to be both objective and creative. There are usually a number of obvious ways in which one should be better serving one's students. For example, if most of a department's majors are preservice teachers then the mathematics major should not be geared towards the needs of those continuing on for doctoral study in mathematics. Such an honest assessment was the start of the current successful St. Olaf major. There are also new approaches that might serve new groups of students, as happened with Lebanon Valley's actuarial science major.

Building support for innovation that engages the faculty. Developing departmental support for change can either start from a broad base, as took place at Mt. Holyoke, or can involve a few key people who convince others of the merit in reform, as happened at the University of Michigan. Those key individuals start instructional reform activities and slowly enlist the participation of others.

Initiating the process of change and experimentation. It is very helpful for faculty to become actively engaged in instructional innovation. Doing new things in the classroom typically causes the instructor to come across to students as a more active teacher which stimulates students to be more active learners in the classroom. Continuing experimentation was the hallmark of most of the institutions visited in this report, even though they already had successful programs.

Developing an environment of faculty involvement in the welfare, academic and otherwise, of their students. Whether it was described as good coaching, caring about one's students, or commitment to seeing students succeed, the critical aspect of all effective programs noted in this report was the connection that faculty made with their students inside and outside classrooms.

Report on Site Visit to Lebanon Valley College

Ivar Stakgold and Alan Tucker
November, 1993

Lebanon Valley College is located 25 miles east of Harrisburg, Pennsylvania and has a little under 1,100 students. Its students overwhelmingly come from the local region—within 100 miles of campus—where it is reasonably well known. Beyond this core drawing region, it has little name recognition. The student body is virtually all white Protestant; the college has a Methodist affiliation, but this is mostly a matter of history. It is a liberal arts college whose students have SAT scores averaging a little over 1000. It has special strength in music. It also has a tradition of attracting a large number of majors in science and mathematics. Over a third of its graduates are in this area and it ranks 3rd among colleges with under 1000 students in the number of students getting PhD's in science/mathematics. Its 70 faculty teach 4 courses a semester. Promotion and tenure is based solely, in most departments, on teaching effectiveness.

An unusual aspect of Lebanon Valley College is that students have been expected to select a major on admission. In turn, recruitment is closely tied to academic departments. Prospective students visiting the campus will spend half their time visiting with faculty and students in their prospective department. The college has a number of "Day at Lebanon Valley College" events when prospective students visit during a weekday and spend time visiting classes. The first day of the site visit, which started at 2pm, was such a day, and one professor mentioned that there were 20 visitors in one of his morning classes. This policy of direct admission into a major has been relaxed by a new president: incoming students now declare an intended major, with the option of an 'open major' for undecided students. The institution's enrollment was being hurt in the late 1980's by the early major requirement, although Lebanon Valley faculty continued to back early majors. Requiring students to choose a major on admission apparently appeals to a limited, although able, group of students (most students who apply are accepted). Admission is determined by the Admissions Office and departments are required to accept all admitted students interested in their major.

In the late 1960's, the Mathematics Department decided to offer an actuarial science major. The program was promoted in the mid- and late-1970's after ERISA legislation required actuarial certification of proper funding of pension programs, thus creating an extra demand for actuaries. LVC appears to be the only small liberal arts college in the country with an actuarial science major. The actuarial option proved very attractive to students who liked mathematics in high school but did not know of any career using it except school teaching. The number of mathematics majors grew from around 30 to close to 90 during the 1970 s, with the actuarial science option getting much of the credit for attracting the additional majors. Over the last 20 years, the Mathematical Sciences Department has produced 6–10% of Lebanon Valley graduates. While many students change in time from the actuarial major to the regular mathematics major, 25% of the mathematics/science graduates from 1975 to 1987 were actuarial science majors. All but one or two found immediate employment in actuary-related positions. Over thirty graduates have gone on to become full Fellows of one of the two actuarial societies.

The six LVC mathematics faculty offer a major in computer science, as well as in mathematics and actuarial science, plus a secondary school teaching option. There is no remedial mathematics taught (students are sent to a local community college for this). Introductory courses in pre-calculus, calculus, statistics, finite math, and mathematics appreciation are moderate size to minimize the number of sections (e.g., 3 sections of Calculus I handle about 110 students); a couple of adjuncts are used to help with pre-calculus and finite math. This allows the six faculty to offer a large number of upper-division courses—20 mathematics courses beyond calculus plus 10 computer science courses. There are also independent study courses, topics courses and problem seminars, which are often taught informally—not a course the faculty get credit for—when only a couple of students enroll.

The mathematics faculty all care very much about their students. They each know virtually all upper-division majors by name and can talk at length about the strengths and weaknesses of each. A theme that came up many times in conversations with faculty was intellectual growth of students (as opposed to learning particular material), despite the content-based actuarial exam. Students also had a strong awareness that intellectual and personal growth was the objective of their college work. They saw the faculty pushing and helping them to grow. Several students commented that they liked the atmosphere and personal attention of a small college—this is the primary reason they chose to come to Lebanon Valley—and they liked the mixture of liberal arts and strong career orientation of the college and the mathematics program. They also like the diversity of course offerings in mathematics. All students the visiting team met seemed to be very happy with their education. They liked the small number of mathematics faculty that led to them taking several courses from most professors. Students noted that they get to know professors more personally and become comfortable around them in a second course.

Here is an example of the faculty interest in students. Two of the freshmen mathematics majors are commuters this year (commuting students are fairly rare). Although the department chair was not teaching the freshman calculus course for mathematics majors, he knew about the two commuters and was worried that they would be isolated. He arranged for the two commuters to use the mathematics conference room every morning in the first semester for an hour or two to have a place on campus to meet and work together on their calculus assignments.

Although the attraction of actuarial profession and a general emphasis of strong career placement lures students to the mathematics major at Lebanon Valley, the mathematics major is quite rigorous and the faculty set reasonably high standards for student performance. This starts in the freshman year, when mathematics majors take a separate course in calculus, meeting 5 times a week instead of the normal 4-credit first-year calculus sequence for non-majors. This course is demanding with a good dose of theory. Students report spending a couple of hours every night on this course, often in study groups. It is a bonding experience for mathematics majors who take pride in meeting the challenge of this course, but about a quarter of the freshmen mathematics majors leave the major during or after this course.

There is a mathematics club that is not very active in mathematical matters, but does organize social events for mathematics majors (and the faculty). It also organizes a high school mathematics quiz bowl for local high schools.

Students seem to work mostly alone after freshman calculus, although in fraternities and sororities, there is a lot of helping one's brothers and sisters. Many students participate in intercollegiate athletics or have a serious musical interest, say 50% of the mathematics majors (and students generally). The time devoted to sports or music seems to work against the mathematics club.

An advantage of the actuarial profession is that the publication by actuarial societies of the names of the candidates—and their addresses—who sit for actuarial exams and their scores allows the faculty member in charge of the actuarial program, Professor Hearsey, to keep track of actuarial graduates. He produces a newsletter for actuarial graduates telling which LVC grads passed which exams along with other information about the college. This link with actuarial graduates is used in recruiting—they are asked to talk to Lebanon Valley College applicants interested in mathematics who live near them. One would suspect that a successful actuary would be an attractive role model to a potential actuarial/mathematics applicant. (The only mathematical role models most students see are faculty.)

The success of the Lebanon Valley mathematics program cannot be ascribed to any one person. The chair in the late 1960's, Professor Bissinger, started the actuarial science program, although it did not take off until decade later. The chair in the 1970's (still on the faculty), Professor Mayer, was the driving force in creating interest and excitement about mathematics at LVC. Professor Hearsey has done a wonderful job directing actuarial efforts. Professor Tousley, the chair from 1984–1994, was very effective in promoting goodwill among students and faculty. Professors Fry, Townsend and others along the way have also been and continue to be important players in this story.

In summary, the Lebanon Valley mathematics program is very successful because it offers the option of a very attractive prospective career as an actuary, that is enhanced by the use of graduates in the actuary profession to help with recruiting. Further, it has faculty who are extremely dedicated to their students while simultaneously maintaining high standards.

Answers to the Twenty Questions

Q1. Curriculum and course syllabi geared to the needs of typical students; mixture of pure and applied courses.

The relatively large number of majors allows LVC to provide a good variety of courses for a school of less than 1000 students: Eight 100–299 level courses and twelve 300–499 courses in mathematics with an additional 4 upper level actuarial science courses (all quite mathematical) and 5 upper level computer science courses. All the upper level courses are offered at least every other year.

There is a good balance of theoretical versus applied courses. Among the upper level courses, abstract algebra, geometry, probability, statistics, real and complex analysis lean towards the theoretical; whereas operations research, numerical computation, applied statistics, and the actuarial sciences courses lean towards the applied.

Q2. Teaching styles and active participation of students in class.

Most teachers use a modified lecture format with a lot of questioning of students by the instructor, as well as students asking questions. There is an expectation that students should be prepared to participate in class discussions—that is, students prepare for class. Prof. Tousley's classes are built mostly around students going to the board to work assigned problems and then leading a discussion of these problems. The personal style of each faculty member is different. Students seemed to like and respect all styles.

As noted above, there is a lot of use of study groups in the freshman calculus course for mathematics majors and it drops off as students become more self-confident about their mathematical skills.

Q3. Requirements of the mathematics major and, if present, secondary mathematics teaching programs.

Students in all the department's majors (mathematics, actuarial science, and computer science) are required to take the year-long freshman calculus for mathematics majors, introduction to computing, linear algebra, and foundations of mathematics. The mathematics majors have to take discrete mathematics, two 1-credit problem-solving seminars, a probability and statistics course, a course in algebra or analysis, an operations research or numerical analysis course, and at least two electives (most take more). The actuarial science majors take courses in the mathematical sciences and actuarial science covering material on the first seven exams of the Society of Actuaries.

Q4. Nature of out-of-class contact between students and faculty.

As noted earlier, faculty are concerned and closely engaged in the work of their students. The education of the students is their lives. Perhaps most notable is Professor Fry who we were told is more often than not helping students late into the night. There are organized events by the Mathematics Club, including picnics, parties, trips.

Q5. Advising of mathematics majors; assistance in career selection, career placement, and selection of graduate schools.

Every student has a faculty advisor whom they see at least once a semester, frequently more often. In time, students have spent enough time out-of-class with all faculty that all faculty become de facto advisors to all mathematics majors. There is a department-focused freshman experience program during the first semester as an orientation to the college and department.

The actuarial science majors need little career advising. The reputation of the LVC actuarial science program plus summer internships make placement easy. There appears to be an excellent placement record outside of actuarial science also. Career advising in the department is supplemented by the efforts of an effective Career Planning and Placement Office.

Q6. Extracurricular activities, e.g., mathematics club.

As noted above, most of the activity of the Mathematics Club is social: an annual picnic held recently at Professor Hearsey's house, a couple of parties a year at Professor Mayer's house, participation in managing the student pub, and other similar activity. This club also has a long history of organizing and providing a mathematics competition for local high school students. The current format has teams from 12 schools on campus for a day in the spring in a quiz bowl format. Except for proof reading and judging by faculty, the contest is completely student run.

Q7. Mathematics Help Center and other walk-in help programs.

This is mostly handled through the College Academic Support Service. Most of the mathematics and computer science help comes from department majors, who also provide tutoring in accounting and a variety of other disciplines. This program provides for walk-in help and for individual tutors. The department also arranges for its majors to tutor local school students when they receive calls for assistance.

Q8. Level of innovation in curriculum and teaching methods.

Professors Townsend and Liu have attended sessions on using calculus software, Hearsey and Townsend participated in sessions of cooperative learning. As described in Q1, the teaching styles are quite traditional, but represent the best of 'old-fashioned' teaching—extensive questioning of students by teachers, students working problems at the blackboard. Likewise, the curriculum is quite traditional but also very broad for a small liberal-arts college. Of course, the actuarial science major, while common at universities 60 years ago, is quite innovative for a small college mathematics program.

Q9. Ways to recruit students to become mathematics majors, through freshman mathematics courses, extracurricular means, recruitment materials.

Essentially all mathematics majors select mathematics or actuarial science as a major before they arrive at Lebanon Valley. As described above, the department is very actively involved in recruitment on campus. There is also extensive off-campus recruitment through writing to high schools and some high school visitation, but mostly through working with the College Admissions Department. Prospective students, whether identified by off-campus contacts or campus visits receive follow-up telephone calls from mathematics faculty. Many mathematics majors are first attracted to LVC by the actuarial science program.

Department retention is significantly enhanced by the curriculum which places all mathematical science students in the special calculus class during their freshman year. This allows and encourages students to get to know other mathematics students and to become part of the mathematics department early in their undergraduate life.

Q10. Support programs for women and/or underrepresented groups.

Women seem well represented in current and past LVC mathematics majors, a little under half of the mathematics and actuarial science majors. The department is conscious of not having a female faculty member although it has apparently tried hard to recruit women. The massive amount of attention that all students receive seems to have been effective at minimizing any potential problems for women in mathematics. LVC has very few African American students generally, and particularly in mathematics.

Q11. Use of technology.

The department has not been especially active with technology for a variety of reasons including heavy loads which get in the way of learning to utilize effectively new technology. Most of the faculty have used some software to enhance presentations and to get students involved with computing. The personal environment makes some of the efficiencies of some technology less relevant. There is a 32-seat classroom equipped with 386 and 486 microcomputers.

Q12. Transfer coordination from two- and four-year colleges.

The Lebanon Valley mathematics department has a carefully worked out arrangement with Harrisburg Area Community College (HACC) which has brought in one or two mathematics students a year over the last few years.

Q13. Facilities for mathematics majors: departmental computing lab, study room, mathematics library.

For a small college, there are extensive microcomputing facilities available, along with terminal connections to a VAX, but no workstations. The department maintains a study room/meeting room with a good set of reference texts, periodicals and course notes. The central library subscribes to 30 technical journals in the mathematical sciences.

Q14. Independent study, senior thesis, summer research opportunities, and internships.

There is a lot of independent study, mostly either tutorials for unoffered courses or for work on honors projects. Many graduates have had business and industrial internships. Most of the computer science students have this opportunity, as have roughly three fourths of the actuarial science students.

Q15. Honors courses and programs.

A significant number of mathematical science students do participate in the College Honors Program. These students must complete a project of some type.

Q16. Special topics and research seminars.

There are two senior seminars on problem solving and famous problems, each 1-credit courses, that are required for all mathematics majors.

Q17. Colloquium series; talks by alumni and industrial mathematicians and managers about uses of mathematics.

The LVC mathematics department have not done much in this direction outside of actuarial recruiters talking about what actuaries do. This fall they hosted their MAA section conference on "Careers in Mathematics." (Professor Fry, who is on the Section Board of Governors, was the organizer of the conference.)

Q18. Formal program of student input to department, such as a student member of the department curriculum committee.

The size and environment naturally leads to a lot of student input, although there is not any formal mechanism.

Q19. Training programs for junior faculty, particularly for non-native speaking faculty.

There is no program and their small size would not suggest the need for an ongoing program, since a new faculty member is hired only about once every 5 years. They do have a process for evaluation and discussion of teaching.

Q20. Special programs and accommodations for evening and part-time students.

No part-time or evening students.

Contact: Professor Bryan Hearsey, hearsey@acad.lvc.edu.

Report on Site Visit to Miami University of Ohio

Bill Hawkins and Ray Schiflett
October 1993

The visiting team met with groups of faculty and students and also saw the Dean of Arts and Science. By the end of the two days, they felt they had a fairly good impression of the Department of Mathematics and Statistics at Miami University.

Our overall impression was positive. Individual faculty or groups of faculty are trying new things and proposing changes. The students were generally enthusiastic about the department and their overall educational experience at Miami University. Moreover, the department faculty are also very concerned about maintaining their research base.

The main body of this report is written as responses to the twenty questions developed by the project advisory committee.

Answers to the Twenty Questions

Q1. Curriculum and course syllabi geared to the needs of typical students; mixture of pure and applied courses.

The curriculum is a strong and somewhat standard set of course offerings. The department serves well both its own majors and the needs of the university as a whole. They have a heavy service component which will get heavier as the new educational plan for the university is implemented.

The department offers a BA and three BS degree programs. In cooperation with the department of Teacher Education, they offer a BS in Mathematics Education. They have a total of 295 declared majors in these programs (out of 12,000 students). The largest (122) is the BS in Education. The department also offers minors in mathematics, statistics, and operations research methods.

The Department offers the MA, MS, and MAT degrees in mathematics, as well as the MS in statistics. They have about 30 graduate students. The MS in mathematics has an Operations Research option and the program generally allows an individually-designed course of study with approval of the department's Graduate Committee.

Students have a good deal of flexibility as they enter the programs in the department. We were impressed with the clever and thoughtful way the calculus and precalculus courses are designed to meet the variety of backgrounds and entry-level skills possessed by the students. This does not mean, however, that there are a plethora of calculus courses. On the contrary, one of the strengths in the department mentioned by the faculty is that, except for the honors calculus courses, all students take the same calculus course. It is interesting that the department has shown that this one course of study approach actually makes the program accessible and attractive to a wider range of students.

In addition to the flexibility for students entering the programs of the department, there is a flexibility in degree requirements. One option permits a student to design his or her own course of study, subject to approval by a three-person faculty committee, with the Undergraduate Committee making final decisions in complicated cases.

Q2. Teaching styles and active participation of students in class.

The information we were able to obtain about teaching styles and methods in the department suggests that most of the courses are well taught in a traditional lecture mode. There are a small number of faculty experimenting with technology and alternatives to lecturing.

Q3. Requirements of the mathematics major and, if present, secondary mathematics teaching programs.

All of the majors mentioned above require the calculus and linear algebra along with 19 to 28 additional semester hours of mathematics or statistics. An interesting addition to the concept of the major at Miami is that students with an interest in mathematics but majoring in other disciplines can still graduate with honors in mathematics.

Q4. Nature of out-of-class contact between students and faculty.

Students and faculty have moderate contact outside of class and office hours. The department holds a social event at the start of the school year for incoming freshmen, upper class majors, and faculty, although it is not always well attended. Some faculty host groups of students at their homes for a meal or social gathering. The students speak highly of the faculty and are pleased with their experience but this appears to be associated with the classroom more than informal relationships. However, some faculty are very involved with students and student organizations and some students seek close mentoring relations with some faculty.

Q5. Advising of mathematics majors; assistance in career selection, career placement, and selection of graduate schools.

The information on advising was varied and somewhat contradictory. At one point, we heard praise about the advising in the department from faculty, but then heard some fairly strong dissatisfaction from some students. The placement office is not thought well of by faculty or students.

Presently, there is a five person departmental advising committee. Before this committee approach, advising was the responsibility of all faculty. The change was made partly in recognition that some faculty did not like and did not do well at advising.

Q6. Extracurricular activities, e.g., mathematics club.

The department has a very active Pi Mu Epsilon chapter. They hold monthly meetings, usually with an invited speaker. They also run conferences and contests. There is an NCTM-affiliated student chapter that conducts Saturday conferences for middle school children. There are special problem seminars held for the Putnam, Mathematics Modeling, and actuarial exams.

Q7. Mathematics Help Center and other walk-in help programs.

Mathematics majors, with supervision from graduate students, assist in evening help sessions for pre-calculus and calculus classes, but the department has no tutorial room available around the clock. Both students and faculty wish that a mathematics room for students were available.

The university has recently instituted a peer tutoring/mentoring program administered by the Office of Student Affairs. Ironically, its creation has served to postpone the fledgling experiment with the Treisman Emerging Scholars program in the department. The peer tutoring program contracts with "at-risk" student—who are viewed as taking "at-risk" courses—and the peer tutors attend the courses in question. There is no program which offers assistance or tutoring beyond the 300 level courses.

General tutoring is handled by the Office of Learning Assistance.

Q8. Level of innovation in curriculum and teaching methods.

There is some experimentation in both the curriculum and the delivery systems. About half of the faculty have tried new things in their classrooms, and about a quarter are continuing to experiment. Several have tried the Harvard Calculus materials, and last year they ran 6 sections using these materials out of 26 sections of calculus. Some faculty are requiring writing in some courses, and a few have tried some form of group work.

The department voted recently to continue and expand the Harvard calculus experiment, but lecture is the dominant mode of instruction. However, since 80% of the calculus students pass with a C or better, they may not feel the need to be innovative here.

Q9. Ways to recruit students to become mathematics majors, through freshman mathematics courses, extracurricular means, recruitment materials.

The department does no formal recruiting of students.

Q10. Support programs for women and/or underrepresented groups.

There are no programs in place to support or encourage minorities or women. They have tried a few things but nothing substantial exists. The department is above the national average for women undergraduate majors (50% female majors) but has almost no representation from the Hispanic or African-American communities.

There have been only six minority mathematics majors in the past 24 years. There are two underrepresented minority mathematics faculty out of forty faculty positions in the department. One has been there for 24 years, the other for about six years. The student enthusiasm for the mathematics major and the student orientation of the university and the department make their poor record with minority mathematics students surprising.

Q11. Use of technology.

The department is actively using technology in many of its courses but not as a departmental policy. The TI-81 is used throughout the precalculus. Faculty have experimented with Derive, ISETL and spreadsheets. Both the university and the department have computer laboratories used by mathematics students.

Q12. Transfer coordination from two- and four-year colleges.

Miami University has two regional campuses, at Hamilton and Middletown, that offer the first two years of a college program. Students can transfer to the Oxford campus after one or two years. The remedial mathematics programs are located at these two campuses.

Q13. Facilities for mathematics majors: departmental computing lab, study room, mathematics library.

The department has a computer lab but no student oriented room or library.

Q14. Independent study, senior thesis, summer research opportunities, and industrial internships.

There is a good deal of undergraduate "research" mathematics done by students which is presented at regional and national meetings. The department has an industrial internship program and an active actuarial program. There are no internships presently. The department encourages students interested in actuarial science to pass at least the first exam before graduating. There are several related courses that such students are advised to take.

Q15. Honors courses and programs.

There are honors sections of calculus and linear algebra. As noted in Q3, students with other majors who do well in their mathematics courses can graduate with Honors in Mathematics.

Q16. Special topics and research seminars.

Many of the department's faculty are research oriented and do hold research seminars. Advanced undergraduate and graduate students are invited to attend.

Q17. Colloquium series; talks by alumni and industrial mathematicians and managers about uses of mathematics.

The department hosts a major conference each year on a mathematical theme. There are student paper sessions at this conference and Miami students contribute several papers at these sessions. The department has alumni, mathematicians outside academia (including actuaries), and academic mathematicians come to campus to give talks in a regularly scheduled colloquium series. The Pi Mu Epsilon chapter also hosts outside speakers.

Q18. Formal program of student input to department, such as a student member of the department curriculum committee.

The department has a very impressive commitment to student input on departmental issues. There is a Student Advisory Board and its members have a voice and vote on almost all issues which come before the faculty.

Q19. Training programs for junior faculty, particularly for non-native speaking faculty.

The department has a week-long orientation for all GA's and a special program to help graduate assistants who are teaching precalculus. They have a TI-81 orientation session and hold weekly meetings. Older faculty mentor younger faculty. The university also sponsors a 'teaching scholars' program for young faculty.

Comments by the Dean of Arts and Sciences

The Dean seemed generally positive about the department. He commented on the importance that the university placed on having students graduate with significant mathematical skills. There is a formal reasoning requirement for all students, which most students satisfy through mathematics. He also was aware of the changes in some of the calculus sections and that students appreciate those changes. He mentioned some of the specific work being done using spreadsheet software to teach courses. He also commented on the excellent and cooperative relationship that exists between the department and the school of education.

The Dean did mention that he is not satisfied with the nature of the evaluation process in the area of teaching. He indicated that the Provost and President wished to develop a broader view of what constitutes scholarship and establish a new standard for promotion and tenure which credits scholarship in the areas of teaching and learning. Presently, most departments require publication of refereed papers. The Dean indicated that he perceives the mathematics faculty are not in favor of such changes and wish to have the emphasis remain on published mathematics research.

Comments from Students

The students were remarkably talented, cheerful, and positive. They had many good things to say about the faculty and department. They discussed their involvement in the department in addition to course work. For instance, they talk about their work in help-sessions they conduct for the department in precalculus and calculus. These meet Monday through Thursday from 6 to 8 p.m. with around 4 graduate students in each of the two rooms. The students stated that there is limited contact with most faculty outside of office hours.

Comments from Other Departments

Our final session was with faculty from cognate departments and also included the Dean of Applied Science. They were uniformly complimentary of the cooperative attitude of the mathematics faculty. The Dean spoke at some length, giving examples of how the department worked to meet the needs of the faculty and students from his school. This was echoed by the faculty from those departments represented.

Contact: Professor David Kullman,
dekullman@miavx1.acs.muohio.edu

Report on Site Visit to Mt. Holyoke College

David Lutzer and Ivar Stakgold
March, 1994

Context

Mt. Holyoke College (MHC) is a 1800-student liberal arts college for women. The mean SAT cumulative score for entering students is 1128 and the graduation rate is about 84%. MHC mathematical sciences faculty teach four courses per year. A mathematics class of forty is seen as extremely large by MHC standards. MHC's mathematical sciences department is called the Department of Mathematics, Statistics, and Computer Science and it offers three separate majors with quite different requirements. Each spring, about twenty-five students graduate with degrees from the department—about twenty in mathematics, five in CS and a few in statistics—and that is about 5% of the graduating class.

MHC is part of a five college consortium (Amherst, Smith, University of Massachusetts at Amherst, and Hampshire College being the other members). One of the major NSF calculus reform projects resulted in a calculus text, *Calculus in Context,* that was co-authored by members of several consortium departments. In addition, some mathematics faculty members teach courses on each others' campuses. The consortium also creates "critical research mass" among the faculties of the five schools and makes valuable contributions to faculty research. The consortium is also intended to enhance the undergraduate curriculum, but our meeting with nine mathematical sciences majors suggests that few students take advantage of their right to take courses at other consortium members. Several times we heard of discussions to insure that certain advanced mathematics courses are scheduled in such a way that students always have access to central mathematics courses in any given semester at some member of the consortium. The level of cooperation made possible by the Five College Consortium is unusual.

What is especially noteworthy in MHC's mathematical sciences department is the level of curricular reform and experimentation being carried out by a very active faculty. The outside world has noticed the energy and expertise of MHC faculty: national foundations have literally poured money into the department—several million dollars over the last six years.

Unusual and Exportable Features at Mount Holyoke College

We saw three major themes at MHC that might be of interest elsewhere. One major theme is "mathematical sciences for everyone". A second is "mathematical sciences in context," and a third is "mathematical sciences as experimental science." We also saw a way to offer mathematics and statistics curricula which are very broadly aimed and yet can provide real depth for some, provided a department is willing/able to stress undergraduate research and honors theses for its best students.

Something important needs to be stressed here. Unlike the situation that one of us saw at the University of Chicago (whose students may be so bright as to make the Chicago curriculum nonexportable), typical MHC students do not enter college with backgrounds and abilities that are far beyond what one can find at many private schools and some large state universities. This leads us to believe that some of the very unusual MHC ideas that we encountered might really be transferable.

Theme I: Mathematical Sciences for Everyone

We saw the "mathematical sciences for everyone" theme played out in two ways. First, we saw unusual levels of involvement by many mathematics department members in the design and teaching of general education courses, including multidisciplinary courses outside of the department that have mathematical components. Second, there is an ongoing attempt by the mathematical sciences department to cut down the rather linear prerequisite structure that is

42

typical of our discipline so that students can enter the mathematical sciences major or minor through many doors.

We saw one example of a very impressive general education course called "QR" (for "Case Studies in Quantitative Reasoning"). It was designed by members of several departments (with significant Sloan foundation support) and is taught by teams of mathematicians, statisticians, and faculty from other departments. Its content seems to be tilted toward social science applications of mathematics and statistics. Section size is small and it is taken by roughly 15% of the student body. Course content can vary from year to year, but the structure remains the same. There are three main case studies in each year's course, all of which involve mathematical and statistical methods and all of which have a nonmathematical context. Students attend one lecture per week and two discussion sections, plus a required three hour computer laboratory. Students write major reports growing out of each of the three sections. A recent QR course was built around the following three topics: "The Salem witchcraft trials: wealth and power in Salem village, 1681–1696"; "SAT scores and predicting GPAs"; and "Modeling population and resources."

Another general education course, Past and Presences in the West (P&P), focuses on the humanities, but one of its six units is always on science or mathematics. In recent years P&P has included a unit on mathematics, either number theory and diophantine equations or geometry. We have less information on that course.

In addition to these multidisciplinary general education courses, the department offers general education courses of its own. We visited a number theory course for freshmen with no mathematical prerequisites. (There is a similar freshman geometry course.) In a class just after the middle of the MHC semester, freshmen with no previous college mathematics were working on the problem "Can one prove the quadratic formula in the field $\mathbb{Z}$ mod p?" After talking about the usual algebraic proof in $\mathbb{R}$, the instructor broke the class up into subgroups and assigned them variations of the problem "What about $3x^2 - 2x - 5 = 0$ in $\mathbb{Z}$ mod 11?" The instructor later described the course as headed toward an algebraic geometrical result that monic polynomials with rational integral solutions have solutions in all moduli. That is quite different from the intellectual level in many "Mathematics for Poets" courses that exist around the nation. Another interesting feature of this course is that it can serve as a suitable prerequisite for more advanced mathematics major courses (see below). A meeting with graduates of this course elicited high praise for it.

In an attempt to allow multiple entry points for the mathematical sciences major and minor, the department has modified its prerequisite structure. For example, students can enter second year and upper division mathematics from many initial courses. Calculus I of some kind (including a special year-long course that combines pre-calculus and calculus) remains a prerequisite for many sophomore and upper level mathematics major courses, but for courses such as Linear Algebra or Discrete Mathematics, any 100 level course (e.g., the number theory course for non-majors described above) will do. A course titled "Laboratory in Mathematical Experimentation" (discussed below) is a fulcrum of the mathematical sciences major, and its prerequisite is calculus I of some kind (including the year-long course that combines pre-calculus and calculus) or the freshman number theory or geometry course. The unusually flexible prerequisite structure extended beyond the freshman and sophomore years. For example, there is a senior course on Lie Groups that has Calculus I and linear algebra as its prerequisites. There is an upper division course in knot theory which students may enter after completing Calculus II (or even after the year-long course that combines pre-calculus and calculus I) or the freshman seminar in number theory or geometry. The department has found that the majority of its majors and minors still enter through the main door (Calculus I, II, III), but we did meet several graduating seniors who began in different courses and took calculus late in their college careers. Because different departmental documents said different things about the prerequisite structure for various courses, we deduced that the entire issue is still in flux.

Theme II:
Mathematical Sciences in Context

The second major theme at MHC is "mathematical sciences in context." Of the traditional introductory courses for mathematics and science majors that department members described to us, most try to teach mathematics in the context of a major applied theme. In particular, most calculus courses at MHC use an applications-based text written by Five College faculty titled *Calculus in Context*, that is part of the calculus reform movement and is available commercially. Repeated visitations to a few very applied topics provide one aspect of the intellectual glue that holds the calculus courses together. Another unifying feature is that computer usage is emphasized throughout the calculus sequence. More important, in the department's view, is the emphasis given to differential equations and techniques of successive approximations in the courses.

While most mathematics faculty are very positive about the course, of the nine upper division mathematics majors with whom we met, several strongly disliked the approach of Calculus in Context, and the majority said that even in retrospect, they were confused about what central theme was meant to organize the course's intentional intermixing of traditional calculus topics and significant application streams. In response to a draft version of this report, the department questioned whether the visitors saw a representative sample of Calculus in Context students. They noted that the majority interviewed either had not taken Calculus in Context at all or had taken only a version of Calculus in Context II that was substantially changed in 1993.

To help us understand how well the mathematics department is meeting the needs of its client departments, we were introduced to two members of the physics department who said that many members of physical science departments are dissatisfied with the new approach to calculus in terms of what typical students in introductory physics courses know about the manipulative skills of calculus. On the other hand, the mathematics department reports that the social and life science departments seem quite pleased with the Calculus in Context approach. As has always been the practice, each calculus instructor chooses the text for his or her section. Some visitors have used traditional texts, while department members have chosen a variety of reform texts. Currently the department schedules one non-Calculus in Context section of Calculus I each term.

Theme III:
Mathematical Sciences as
Experimental Science

The third major theme at MHC is the role of experimentation, student participation, and active learning in mathematics and statistics courses. That experimentation theme, rather than the traditional "applied vs. pure" dichotomy, is the right way to understand what MHC's mathematics program is about. The theme of student participation, mathematical experimentation, and written communication about mathematics cuts across every course that we heard discussed, both pure and applied. As part of this theme, the department aims to shift mathematics courses away from the lecture format and toward participatory learning by students, with heavy emphasis upon computer experimentation. Students and faculty members told us that lecturing is

still a part of MHC pedagogy, but whenever possible other pedagogies are employed. We visited two courses and in each we saw evidence of independent student work, both computer-based and with pencil and paper.

A four credit mathematics experimentation laboratory is required of all mathematics majors, and is normally taken during the sophomore year. Students in the laboratory that we visited were required to submit six 10-page reports on mathematical experiments dealing with such topics as iteration of linear functions; iteration of quadratic functions; numerical integration, starting with Riemann sums; Euclid's algorithm; and infinite series (Taylor and Fourier series). The extent of instructor involvement in these projects decreases as the semester progresses. At first, students write reports based upon structured outlines provided by the professor, but by the end of the semester students are much more on their own in terms of posing questions and choosing their own experiments to address the questions. In each of the topic areas of the course, students must write reports about their work, explaining why this or that experiment was tried and why this or that conclusion seems warranted. According to faculty members, this orients students toward explaining why things work, right from the start of their mathematical studies, and that makes the later study and writing of proofs easier for students. Upper-class students who were interviewed had only wonderful things to say about this course. They all said that writing the papers forced them to get deeply engaged in understanding mathematics in a way that no regular course ever did.

Study in Depth and Later Studies

The MHC mathematical sciences curriculum does not fit comfortably with the latest CUPM recommendations on the major in mathematics, in that MHC does not require any year-long sequence at the upper division level. Students are required to take Real Analysis I and Modern Algebra I, and might follow them up with an elective that builds on and extends these courses, but some choose not to do so. Several MHC faculty members commented that students who want to go deeper into a subject frequently request independent study courses with faculty members, and there is a lot of independent study work going on at MHC. The students with whom we met made it clear that it was well-known and common practice to seek independent study help from the faculty.

It is interesting to note that by the time MHC students are seniors, both the faculty and the students themselves are pleased with all students' ability to understand and write mathematical proofs. When questioned on this issue, faculty members tended to ascribe this success to the fact that, throughout all four years of undergraduate mathematics study, students are asked to give written explanations of their thinking in English, starting even before they are asked to think about proofs. This is certainly an important aspect of the MHC program that could be very exportable and deserves further investigation. Some faculty said that there was a noticeable increase in students ability to ask good questions and attack proofs since the introduction of the sophomore laboratory course.

The department does not see the minimal requirements of its mathematics major as adequate preparation for graduate studies in mathematics. The department's solution is a combination of undergraduate research and honors thesis work for its strong students planning advanced study in mathematics. MHC runs an REU program with NSF funding during the summer. Under NSF rules, only a few of MHC's own majors can participate in MHC's program; more to the point, the department encourages its students to go to other REU programs around the nation between their junior and senior years, where possible. (Foreign students, it appears, cannot receive REU support from any NSF program, and that is a real problem since MHC has many foreign mathematical sciences majors. As a result, MHC finds private money to support these foreign students to participate in REU experiences at MHC.) After completing REU experiences, students are urged to write honors theses, often growing out of their summer research experiences.

Conclusion

The mathematics program at MHC is alive and well. It features an enthusiastic, energetic, and highly active faculty and has taken to heart MAA's call for a broadened definition of mathematical scholarship, and for activity in curricular experimentation. MHC mathematics students are also enthusiastic about their teachers and about the department's program. We believe this is a very successful program having many features that could be emulated elsewhere.

Answers to the Twenty Questions

(Questions that are answered fully above are skipped.)

Q3. Requirements of the mathematics major and, if present, secondary mathematics teaching programs.

Requirements for mathematics major: Calculus III, Linear Algebra, Mathematical Experimentation Laboratory course, Analysis I, Algebra I, and three more courses at the upper division level. Each of the department's courses is a four hour course.

It is possible to be certified for mathematics teaching at MHC. A specific mathematics course (Math 397) is required, plus a number of courses taught by the Psychology and Education Departments.

Q4. Nature of out-of-class contact between students and faculty.

See above. Students find faculty members very accessible and readily available for independent study projects.

Q5. Advising of mathematics majors; assistance in career selection, career placement, and selection of graduate schools.

Students felt confident that faculty members would know about these things, and saw faculty as the first place to go for information. The department is able to describe where its alumni have gone for jobs and what they are now doing.

Q6. Extracurricular activities, e.g., mathematics club.

There is a moderately active mathematics club.

Q7. Mathematics Help Center and other walk-in help programs.

Upper division students are employed in tutoring services for Calculus I and II students. These sessions run every evening and are well attended.

Q9. Ways to recruit students to become mathematics majors, through freshman mathematics courses, extracurricular means, recruitment materials.

The department's faculty visits magnet high school programs to talk to prospective students about mathematics and science at MHC. There does not seem to be much out-of-class recruiting of MHC students. The department faculty arranges for majors to put on skits for freshmen during orientation, and holds teas for prospective majors. However, the department's theory of multiple entry points into the major is an innovative idea and we did meet two students who entered through non-traditional course sequences, one starting in the QR course mentioned above.

Q10. Support programs for women and/or underrepresented groups.

Very strong. This is a women's college, and it does seem to have good ethnic diversity among its students. MHC has a special program for older women who want to return to complete their college educations, or perhaps try college for the first time. It is called the "Frances Perkins Program." In this program, older women attend classes as part-time students for a few years, and then join the student body full time for their last year or two.

Q11. Use of technology.

Very good. There are computers all over the place, in almost every classroom. The faculty is very proud of its self-designed software. Extensive computer use in mathematics courses starts in the freshman year.

Q12. Transfer coordination from two- and four-year colleges.

There is some transfer activity at MHC. The Frances Perkins program (see Q10) for older students is the closest thing to an organized program for transfer students at MHC, and it seems to be very successful. At any given moment, there are perhaps 100 Frances Perkins Scholars at MHC and most have transferred from community colleges. Given that the total student population at MHC is only 1800, this is not negligible.

Q13. Facilities for mathematics majors: departmental computing lab, study room, mathematics library.

Good computing facilities, including several computer laboratories with 486 PCs and workstations. In addition, there is a reading room for mathematical sciences majors to use. Q14. Independent study, senior thesis, summer research opportunities, and industrial internships.

REU experiences at MHC and in other mathematics departments, followed by senior honors theses, are an integral part of their plan for adding mathematical depth to the programs of the strongest students. The nine students with whom we met said that the faculty was a good source of industrial internships, and the faculty said that they work closely with the college placement office on such things.

Q16. Special topic and research seminars.

The department schedules certain special topics seminars on a two year rotating basis. It does not seem to be important which of these upper division seminars a given student takes. Topics include knot theory, polyhedral differential geometry, symmetry and group theory in physics, and Lie groups. Prerequisites are kept to a minimum in about half of the special seminars. In addition, students frequently study special topics through independent study courses.

Q17. Colloquium series; talks by alumni and industrial mathematicians and managers about uses of mathematics.

We heard about an alumnae panel on careers run every other year in which former students return to talk about mathematical opportunities in business and industry and their own experiences after graduating from Mount Holyoke. In addition, the department schedules five or six talks per year for its students by visiting mathematicians and statisticians from academia, government, and industry. This seems to us to be quite a high level of activity in this regard.

Q18. Formal program of student input to department, such as a student member of the department curriculum committee.

Students feel closely involved in the department and feel they are listened to about matters of concern to them. In addition, students meet with and evaluate candidates for faculty positions—and the department takes their input seriously.

Q19. Training programs for junior faculty, particularly for non-native speaking faculty.

We did not see any of this. All of the department's faculty members are native speakers.

Q20. Special programs and accommodations for evening and part-time students.

There are essentially no part-time students, except those in MHC's Frances Perkins program (see Q10). There are no evening students.

Contact: Prof. Harriet Pollatsek,
hpollats@mhc.mtholyoke.edu

Report on Site Visit to St. Olaf College

Linda Boyd and James Leitzel
November, 1993

St. Olaf College enrolls about 2900 students. The graduating class, usually around 700, includes 75 to 100 mathematics majors, or 10 to 15 percent of the graduates. St. Olaf enjoys this success for a variety of reasons. In our brief visit to the campus we were able to discern a few of them.

In the late sixties and early seventies, the program was viewed primarily as providing service for majors in the physical sciences. The department was graduating between 30 and 35 majors a year. The program at St. Olaf was radically changed in the mid-seventies. A conscious decision was made by the department to develop a major program that would be accessible to any student entering St. Olaf College. A "contract major" was developed and students were encouraged to think in terms of a double major, mathematics with some other subject. That second choice ranges widely over the majors offered by the College—music, biology, philosophy, economics, as well as the physical sciences.

The department has "concentrations" in statistics and computer science which are available to any student in the College regardless of choice of major. For an individual student, the "contract" is carefully worked through by the student in consultation with a faculty advisor. Even though majors may enter the program with other career options in mind, about 20% to 30% of graduates do go to graduate school. During the period 1978–88, St. Olaf College ranked 5th in the nation among liberal arts colleges in the production of PhD mathematicians.

The "contract major" for students has direct impact on the program. Emphasis is placed on the quality of teaching, especially in entry level courses. A variety of courses are offered that will attract students at an early stage in their study and students are placed appropriately so that they experience early success. Hence, at St. Olaf the calculus sequence is only two semesters. Multivariable calculus is not required (but is taken by almost all students later in their study). Instead, a course in linear algebra, from the structural view, is the required third semester experience.

This course in linear algebra, taught with a strong component of technology, is another key ingredient in encouraging students to major in mathematics.

To receive a major at St. Olaf, students must complete at least seven courses beyond the three just mentioned. There is a 'core' of four courses, including a course in advanced calculus and a course in abstract algebra, one course from applied mathematics, and one course from a list of higher level theory courses. After linear algebra, students have a wide variety of courses to choose among. They must discuss their contracts with faculty members before submitting them to the department. This provides a sense of ownership and involvement by the students in developing their mathematical education and contributes to the satisfaction they feel with the major.

The College is selective, but first year students look pretty much like first year students elsewhere. The placement procedures in the department are one key ingredient in enabling students to succeed. Several factors are accounted for in that procedure and students receive individual advice as to where they should begin their mathematics study. For the 1993–94 academic year, 61% of the 750 incoming students were recommended for Calculus I. The department estimates that by the time of graduation, about 85% of an incoming class will have a calculus experience. And it is in that first semester calculus class that a great deal of attention is showered on students. First, because of the placement process, the faculty recognize and respect each student as having the ability to complete the course successfully.

Early in the term, all calculus students are invited to a departmental ice cream social to get to know the faculty and their peers in a different setting. In December, there is a colloquium titled "To Be or Not to Be" which addresses the advantages of becoming a mathematics major at St. Olaf. Classrooms are clustered in one area of the building in close proximity to the computer laboratory. The hallway has many bulletin boards addressing a variety of items—faculty pictures, career information, history of graduates, mathematics competitions (problems and solutions), activities of the MAA Student Chapter, to indicate a few. Students just beginning their studies see actively engaged upper-class students and mingle with them before and after class. Students know that mathematics is good at St. Olaf. This critical mass of actively engaged undergraduate students draws more majors into the program.

Students attribute the success of the program and why it works so very well to the nature of the faculty. There is no question that the faculty are absolutely key in the endeavor. They are dedicated and committed to providing a stimulating, challenging, and supportive atmosphere for student learning. Faculty interact almost continuously on matters of teaching and learning. There is a weekly seminar that addresses issues in education. One week it may be discussion on use of technology in assessing student learning, another time discussion may turn to issues of women in the discipline. The demographics of St. Olaf are changing in that now about 60% of the undergraduate students are female. The department is taking this shift seriously and asking questions about how best to meet the new challenges.

An overriding impression of the department is that it is never static. There is always discussion of ways to improve course content and instruction, attention to course offerings, appropriate and widespread use of technology, and assessment procedures. Most of the course development activity (and the department has been extremely successful in receiving grant awards) comes from individual faculty wishing to try some innovative approach to a topic. These efforts receive support from all members of the department as an important contribution to making the department grow and become better. At present, technology is well utilized in the teaching and learning of mathematics by all faculty. Each of the classrooms has a computer and display unit. With support from FIPSE, many members of the faculty are currently engaged in writing mathematical projects utilizing the computer for upper division courses.

The College will be instituting a new set of basic requirements within the next year. The department is actively planning a series of offerings that will attract students who are 'calculus- ready' but, because of choice of major, may not enroll in calculus. The faculty wish to provide a challenging and exciting opportunity for these students to have exposure to mathematical ideas (and thus be encouraged to sample more courses from the department).

All students at St. Olaf have e-mail accounts. The faculty are also networked. Thus not only is much department

business and consensus building done through electronic means, it also serves to extend the faculty advising function. Faculty are always available to students in their offices, by phone, or by e-mail.

The faculty make great commitment to students and to the program. There is a weekly colloquium where speakers are told to make the presentation accessible to students. Overall, the faculty have a mutual expectation of active participation in all activities. These range from the weekly colloquia, to the fall Turkey Roast and the spring Pig Roast (organized by the MAA Student Chapter) to the faculty and student musical performance late in the year. Professionally, many faculty are active in the MAA, AMS or NCTM. Since these are the faculty expectations, recognition and rewards for faculty are built upon professional activity that includes writing research papers, writing grants, developing a new area of mathematical expertise, curriculum innovation and implementation, but with the expectation that these MUST be communicated to the larger mathematical community through publications, lectures, or presentations at professional meetings. St. Olaf enjoys its success because the faculty enjoy and respect one another and collectively they "do it all."

Because the mathematics program at St. Olaf includes so many excellent components, it is difficult to single out a few for special comment. The seven components we have decided to highlight were selected because they seem so critical to the program's success and because they offer models that might be replicated at other institutions.

Placement

The incoming student's introduction to the mathematics department is through the placement procedure. This first exposure to the department sets the tone for the experience to come. The placement program was begun in the late 1960's to provide guidance to students in selecting an appropriate mathematics course and is administered by a member of the mathematics faculty who is given a half-course teaching credit for serving as director. In June she sends a letter to all new students welcoming them to St. Olaf and encouraging them to consider enrolling in a mathematics course in the fall. Students are informed that they must take one of two placement tests. They are given advice about selecting a test and preparing for the test. An enclosure contains sample problems.

The placement examinations are administered early during Week One—a week of orientation, department information sessions, placement testing, and registration that immediately precedes the beginning of fall semester. The Advanced Placement Examination consists of a test from the MAA Placement Test Program together with a 25-question test covering topics in the first semester calculus course and is required of students who have had at least one semester of calculus and wish to be considered for advanced placement. The Regular Placement Examination consists of two MAA tests and is required of all other students. In addition to the test questions, students are asked to answer questions about their mathematics backgrounds and their plans and motivation for taking further mathematics courses. The test scores and admissions information are used in regression equations constructed and fine tuned by members of the mathematics department. From these equations one of eleven placement codes is produced and the code along with a letter explaining the recommendations is sent to each student.

The placement information provides guidance for the students to make an informed decision; they are not required to follow the advice. The Director of Mathematics Placement and other members of the department are available during Week One to meet with students and discuss their placement decisions.

The initial effort involved in mathematics placement at St. Olaf is rewarded by the number of students who successfully complete their initial courses in mathematics. Over 90% of the students who initially enroll in a calculus course complete a semester of calculus; and of these, over 90% receive a grade of C− or above.

Student Environment

Because such care is given in advising students to choose an appropriate first mathematics course, the faculty expect students to succeed and that expectation is conveyed to the students. At the core of the student environment is the expectation of success. That expectation is supported in several ways. First of all, the course structure is designed to allow students to develop mathematical power and confidence.

Another is the recent development of two new courses. After determining that the precalculus course was not providing the proper background or motivation for marginally prepared students to succeed in Calculus I, Calculus with Algebra I, II were developed. Upon completion of these

courses students have covered the material previously included in precalculus and Calculus I. The advantage of this sequence is that the precalculus material is introduced as needed for the development of the calculus. The courses meet five times a week with much more time spent on vocabulary, graphs, and writing than in the previous precalculus course. The course sequence is being offered for the second year, and the faculty is enthusiastic about the results. They report that students are much more willing to attack and persist in solving multi-step word problems. Students worked together more, attended help sessions in greater numbers, were more assertive in class, and earned higher grades.

The student environment is also enhanced by the Academic Support Center (ASC). The ASC provides private tutoring and Mathematics Clinics for many of the students enrolled in mathematics courses. The tutors are recommended and approved by the mathematics faculty and are trained by the mathematics staff member of the ASC. Clinics are available for students enrolled in calculus and statistics courses on a drop-in basis Sunday through Thursday from 7:30 to 9 p.m.

Though these formal support systems make valuable contributions to the student environment, the informal support systems are most important. The students we questioned all indicated that the faculty were always available—they were usually in their offices with the doors open. The large number of mathematics majors and minors is also vitally important. They serve as tutors or paper graders; computer assistants and system programmers; officers and members of the MAA Student Chapter; and participants in mathematics competitions, the Problem Group, the Budapest Semester, the Colloquium Series, and the Mathematics Practicum. The environment provided for the St. Olaf freshman is one in which mathematics is not viewed as an elite discipline for the few, but a lively and accessible subject appropriate for all. The expectation is not only that students will take mathematics and do well in it, but also that they will love mathematics.

The environment also includes a large number of social events. The students not only meet with other students and faculty in connection with their coursework, but also at a number of social events organized by the MAA Student Chapter and the faculty. Some of the events are scheduled at a time when parents are on campus so that they can attend.

Faculty Environment

As with the positive environment for students, there is an environment for faculty that is exciting, challenging, and supportive. A superb library, modern computer equipment, a teaching-learning center, a strong sabbatical program, resources for educational research, and grants for faculty development all support faculty growth in teaching and in professional activity. The goal of the mathematics faculty is to teach students, not just to teach mathematics. That teaching respects both students and mathematics: the faculty insist on high standards and they actively help students meet those standards. Department members are expected and encouraged to teach a wide variety of subjects, ranging across half or more of the courses in the curriculum. Each faculty member teaches six courses a year.

New faculty are quickly assimilated into the Department. They are assigned mentors and meet with them on a regular basis to discuss teaching, to obtain help in planning lessons or tests, and in general to find out about the workings of the Department. One new faculty member told us that she was quickly made to feel included. She suggested that pictures of the faculty be displayed on one of the bulletin boards and that a Handbook for Mathematics Majors be prepared. Her suggestions were enthusiastically accepted, and she was given the responsibility of carrying out the projects.

There are three areas of evaluation for tenure and promotion: effective teaching, significant professional activity, and other contributions to the purposes of the college. These are in descending area of priority, with the third area distinctly subordinate to the first two. Public professional activity includes all published works (texts, research papers, reviews, expository articles, classroom notes), presentations at meetings and at other institutions, leadership in professional organizations, arranging professional workshops, and consulting for industry or academic institutions. Some mathematical research is required for tenure, but it is not demanded as a sine qua non for promotion to professor.

A key difference between the mathematics department at St. Olaf and a department in most colleges or universities is that great efforts are made to operate as a team rather than as a collection of individual specialists. Each individual's professional activity stimulates others in the department. Students benefit both from the results of such collaboration and by observing and often participating in the process: it helps them learn to function as a team in their mathematical undertakings. When we asked one faculty member about the committee structure within the department, we were told that none existed. Individual members have ideas; they present these ideas to the department and work to build consensus. Friction is avoided. He noted, "We value our personal relationships so much that we will not move forward without consensus."

Faculty Staffing

The college gives primary emphasis to effective instruction and that is certainly true within the mathematics department. Many of the faculty commented to us that one of the main reasons the department is so strong is because they have been so careful to place good teaching above every other consideration in the hiring and firing process . They said this has been very difficult and many excellent mathematicians who were good friends and colleagues were not granted tenure. Evaluation of teaching is very extensive and involves evaluation by peers, current students and previous students (including alumni).

Recruitment

The Department actively recruits students through a variety of activities. Each year a Math Day for high school students is held. A visiting master teacher from a local high school teaches a 2/3 load at the College. These teachers form strong attachments to the college and the mathematics department and naturally encourage their best students to attend St. Olaf. Several of the grants that department members administer involve high schools. A current NSF grant has established a network for high school teachers using *Geometer's Sketch Pad*. The best form of recruitment is probably the satisfied St. Olaf mathematics students, a number of whom become high school teachers. They are quick to sing the praises of the program and encourage their friends to attend.

Critical Mass

The students involved in mathematics at St. Olaf form a critical mass.

- There are more than 200 upper-class mathematics majors in this college of 2900 students.

- 85% of St. Olaf students take calculus.

- 12% of St. Olaf graduates are mathematics majors; a little over half of St. Olaf mathematics majors are women.

- Over 40 students are employed as tutors or paper graders.

Because this critical mass exists many things happen very naturally. An example of a natural effect of this critical mass was observed while we were talking with three students in a computer laboratory. Two of the students were mathematics majors enrolled in a linear algebra course and the third was a calculus student undecided about a major. The two mathematics majors assured her that she would definitely become a mathematics major—it was the thing to do. That kind of enthusiasm is infectious for students and for faculty. At the beginning of our visit we asked, "What makes the department work so well?" By the end of the visit we realized there was no single reason. Having a dedicated team of faculty members who are rewarded for excellent teaching attracts students who attract other students who in turn support the faculty and their peers.

Contact: Prof. Arnold Ostebee, ostebee@stolaf.edu.

Report on Site Visit to Seattle Central Community College

Harvey Keynes and William Lucas
November 1993

The visit included: (1) visits to eight classes by each of us separately, (2) discussions with all of the 11 full time faculty and several of the 15 part time members, (3) a session with top administrators, (4) meetings with counselors and other science faculty, (5) a session and reception with some dozen students— mostly former students who have successfully gone on to other universities, and (6) a group lunch with seven faculty. It was a highly efficient and interesting visit. The hosts were most cordial, very cooperative, gave freely of their time, and arranged a good schedule.

All of the classes we visited were highly interactive and group oriented. The style varied from one teacher to another. One taught in a very energetic lecture format where he continually fired questions at particular students and kept the full class most alert. Some had students working in groups of about four with only occasional times when the teacher addressed the full class. Some more advanced classes were entirely in a computer lab format. Calculators were used freely in many of the classes. Seattle Central Community College is blessed to have a collection of excellent, energetic, and dedicated teachers, and this is the primary reason for their success. Most students were active, felt they could freely seek help, and clearly appreciated their teachers' efforts. Our visits included classes at all levels from remedial to differential equations. Help for students beyond the classes and their particular teachers was available in at least two formats and rooms during many hours each day.

SCCC has been extremely successful in maintaining very high quality courses and in preparing their students to succeed in future mathematics and related courses, both at SCCC and in the major universities in the Seattle area (this was confirmed by faculty at the University of Washington). Students transfer with relative ease into neighboring institutions like the University of Washington (UW). Students at times go from UW to SCCC or back and forth between the two schools and do not find the UW courses more difficult. Some students claim there is much better teaching at SCCC, as well as only about half the tuition costs. At times, students even attend SCCC and UW simultaneously.

SCCC students appear to be extremely well prepared to continue studying mathematics when they transfer to a four-year college or university. They are not in any way restricted in their future academic pursuits. Our meeting and reception with some dozen students (out of about 30 invited) confirmed the above assertions about course quality and excellent preparation at SCCC. These students indicated a strong loyalty to their SCCC teachers, expressed a keen appreciation for their teachers' efforts, and realized they had had better than the typical experience. A few of the invited students who could not attend the meeting sent letters or telephoned in their support of the SCCC mathematics program. A testimonial to SCCC's effectiveness is departmental surveys showing that many SCCC students report taking more mathematics than they originally intended upon entrance to SCCC.

Much of the overall success at SCCC appears to be due to a very extensive effort on proper placement, substantial academic and career advising, and continual monitoring of the student's progress. The one math/science counselor whom we spent some time with (Wadiyah Nelson) appeared most capable, described their advising system in detail, and was compiling extensive statistical data to support their activities. Although many transfer students today at two-year and four-year institutions have serious transfer problems, SCCC seems to have minimized this by means of counseling and cooperation with the nearby schools. There is a statewide transfer guide spelling out course equivalencies at all Washington public two-year and four-year institutions.

A contributing reason for so many successful students at SCCC may be the way most classes are scheduled. Many CC's have classes that meet only twice a week, such as two afternoons or evenings per week. Some CC's even have "double courses" (six or more credits) that meet only two times per week. At SCCC, most regular day courses meet for one hour per day on five days per week. Their *express courses*, which proceed at twice the normal rate, meet for two hours per day five days per week. In their

52

express courses, which combine two courses concurrently that are normally taken sequentially, a student with very minimal background can be prepared for calculus in three quarters. Their success both in giving students a richer, more integrated mathematics experience and in preparing many students to succeed in calculus in such a short period could provide a good model for others to follow.

A visitor is greatly impressed with the high energy level of the mathematics faculty and how well they have worked together. They have devoted much additional time developing new approaches in teaching and uses of technology. They have run seminars for each other and taken part in regional and national programs in areas like calculus reform and use of computers. They had a NSF ILI leadership grant for using calculators in beginning algebra.

The college's administrations seemed well aware that they had a good thing in mathematics. (We met with the president, vice president for instruction, and associate dean for science and mathematics.) They have been extremely supportive, given budget constraints, in terms of purchasing equipment, providing rooms, giving some teaching relief, allocating travel funds for some relevant meetings, and in support for the counseling activities. Nevertheless, this has been mostly a "boot strap" type of operation caused mostly by the foresight and efforts of the mathematics faculty. Many of their accomplishments would likely have been achieved more rapidly if additional funding were available.

The ethnic mix of students at SCCC seems to roughly reflect the proportions in the Seattle area. This is a significant achievement given the common overrepresentation of Asian-Americans in most West Coast institutions. Like most CC's they have large numbers of older students and those reentering college. The average student age is about 30. The proportion of women students was large. Many students were oriented more toward vocational training.

SCCC has both academic and vocational sorts of programs. They are organizationally connected with Seattle North CC, Seattle South CC, and Seattle Vocational Institute; and they have a common Catalog that is divided into separate sections for each. Students in vocational courses at SCCC often took regular mathematics courses as part of their programs, and they seemed well served by the mathematics faculty. On the other hand, there appeared to be little unusual in the design of more specialized courses for vocational students per se. Similarly, there were some attempts at offering non-traditional courses or "mathematics for liberal arts" in non-standard formats, but this is not a major activity. These two directions may provide opportunities for

the mathematics faculty in the future.

However, the faculty is currently too stretched to undertake such new developments. Moreover, to do this properly may require additional faculty with either more advanced degrees or more specialized training in some of the less traditional subjects in the mathematical sciences.

SCCC has several outreach efforts to the Seattle high schools and others. Special activities of this sort take place in the summer. However, there seems to be less than full interest in this from the side of the Seattle School District.

One could well conclude from the opinions expressed during our visit that better teaching takes place at SCCC than for similar courses at UW. In a time when many of our large research universities are re-examining their mission and their need for more effective teaching, these two institutions could serve as a good case study. Although the extensive relations between SCCC and UW are mostly of an informal nature, these schools do have a good deal of data that could be useful for such an investigation.

Contact: Janet Ray, jray@guest.nwnet.edu

Report on Site Visit to Southern University

Linda Boyd and William Hawkins
February, 1995

Introduction

Southern University was established in 1879 to provide undergraduate education to "persons of color." Presently the Southern University System consists of Southern University and Agricultural and Mechanical College at Baton Rouge, Southern University at New Orleans, and Shreveport-Bossier City Campus of Southern University at Shreveport. The site visit was made to Southern University and A & M College campus which has 8,000 undergraduates and 2,000 graduate students.

The Mathematics Department is located within the College of Science. The department has 39 full-time faculty members and 4 part-time faculty members. The Department offers both BS and MS degrees in Mathematics. There are options for the BS in both pure and applied mathematics, with the majority leaning toward applied mathematics. The department also has a new BS in Actuarial Science program awaiting approval.

The mathematics department has an extensive committee system, overseen by the department Executive Committee. There are standing committees for calculus, precalculus, statistics, the Mathematics Festival, Pi Mu Epsilon, comprehensive examinations, the Leroy R. Posey Seminar, teacher education, graduate programs, the Mathematics Club, developmental education, public relations, and scholarships. Though this structure appears very rigid, department members appear to work together in very informal ways as well. For example, several members began exploring ways to incorporate technology. They worked individually and in small groups using a variety of approaches. From this exploration it is expected that a department policy on calculator use will be adopted within the next year.

Success with Students

Students interviewed consistently spoke of the faculty as caring people who were genuinely concerned for their progress. The classroom atmosphere created by the faculty allows students to freely ask questions and voice their opinions. In one calculus class that was visited, the students were actively involved during the entire period. Informal groupings of students helped each other while the teacher presented problems and helped groups when they called upon her. These classroom experiences carry over into informal study sessions in the mathematics lab and dormitories.

When interviewed, students always mentioned that their instructors were available to help them. The office area is near the classrooms and students take advantage of the close proximity and open door policy. As well as informal office visits, the mathematics majors are required to visit their advisors twice each semester and go over the course of study. Emphasis is placed on designing an appropriate plan of study to meet future career plans.

Students all mentioned the strong support they get from the mathematics faculty and that their problems and concerns were not only listened to but acted on. Many were very complimentary of the Department Chair. She is extremely responsive to student needs.

It appears that the day-to-day activities that make the Southern University mathematics program work for its students are not that different from those at the other schools visited. They all seem to contain faculty devoted to teaching and supporting their students. The faculty is always seeking ways to improve its courses and programs.

One could not help but notice at Southern a special feeling of pride in the mathematics program and a strong sense of family and of history. A list of mathematics majors dating back to 1942 is carefully and lovingly maintained.

Calculus and the Reform Movement

The Department is very involved in reform in the teaching of mathematics, especially of precalculus and calculus. At the present time, ten of the eighteen sections of calculus are being taught using the Harvard Consortium materials and methods. There are plans to adopt the Harvard Consortium text in the fall of 1996 in all classes of calculus. During the 1995–96 school year, the Harvard Consortium Pre-Calculus text is to be piloted. Teachers and students currently use a variety of technology and software, including graphing calculators, *Derive*, and *Mathematica*.

The impact of the reform movement can be seen in the way students study mathematics, both in class and on their own. In addition to changing student behaviors and attitudes, the reform movement has served as a catalyst for revitalization of the faculty. Faculty members have worked together to learn to use technology effectively and to train themselves in teaching methods involving cooperative learning and Socratic techniques. Several faculty have attended workshops and special programs and obtained funding from outside sources in order to provide students at Southern with both appropriate technology and curriculum.

One example of this is the "Collegiate Curriculum Reform and Community Action" (CCRCA) project. The project involves four members of the department and is supported by the National Science Foundation and Hewlett Packard. Through the program, HP48G and HP48GX calculators are provided for these faculty members and for loan to calculus students. Projection screens are also provided for faculty use in these classes. The faculty team received training in the use of the HP calculators during the summer of 1994 and gave four workshops for students and faculty during the school year. Another example of faculty initiative is a TI graphing calculator project spearheaded by a faculty member and funded by the National Security Agency.

Placement

The mathematics portion of the ACT is used to determine the point of entry into the placement test battery. Students take one of three MAA placement tests—Arithmetic and Skills, Advanced Algebra, or Calculus Readiness. Students with high scores on the Arithmetic and Skills test will then take the Advanced Algebra test and students with high scores on the Advanced Algebra test will then take the Calculus Readiness test. Records of test scores and placements are kept and used for evaluation and modification of the placement system.

Outreach in Schools and Recruiting

Several pre-college programs are offered by Southern. These programs assist, challenge, and stimulate high school students who anticipate enrolling at SU. One of these programs, Upward Bound, offers an opportunity for high school students who have displayed high academic potential to receive assistance in increasing their academic performance. The participants take part in structured scholastic activities that are presented through dynamic teaching methods. The program conducts classes on Saturdays during the school year and Monday through Friday in the summer.

The Department sponsors each year on the Southern campus a mathematics festival for high school students. Last year 658 students of all races competed on various levels— from seventh graders to high school seniors. The competition areas are general mathematics, algebra, geometry, advanced mathematics, and calculus. While the students are being tested, several workshops are held for their teachers. A general session is then held with a lecture given by a Southern faculty member or a visiting lecturer.

Answers to the Twenty Questions

Q1. Curriculum and course syllabi geared to the needs of typical students; variety of offerings and mixture of theory and applied courses.

The positive response from students indicates that the mathematics program is serving the needs of Southern students with mathematical interests. Developmental and pre-college level courses are offered to students who need them but do not count toward the mathematics major. Students then enter the senior division. The applied option in the BS major seems well suited to the many students who are interested in professional careers using mathematics.

Q2. Quality of instruction; active participation of students in the learning process.

The students feel the department is very responsive to undergraduate and graduate students, They think it is one of the best departments on campus, much better than the 5-year engineering program. Teachers adjust their method of instruction to the abilities and interests of the students.

Q3. Requirements of the mathematics major and, if present, secondary mathematics teacher preparation program.

The major requires 40 semester-hours in mathematics at the calculus level or above, including a course in geometry; 30 hours of approved electives; ten hours of French or German; 3 hours of computer science; 12 hours of natural science, including a two-course sequence and both biological and physical sciences. The minor requires 24 hours of courses to include the first two semesters of calculus and the first course in linear algebra. To graduate, students must pass a writing exam and a departmental comprehensive which covers linear algebra, calculus, advanced calculus, differential equations and modern algebra in addition to options based on individual programs.

The department does not offer a secondary mathematics teacher option at the bachelor's level since secondary teachers receive a degree in education with required mathematics courses—college algebra, trigonometry, statistics, Calculus I–III, two geometry courses. The number pursuing this career path is small. If a student earns a BS in mathematics, then passing the National Teaching Examination will give him/her a temporary certificate for one year. There is a Masters in Teaching in Mathematics graduate program that stresses ideas in the NCTM *Standards* and related school mathematics reform.

Q4. Nature of out-of-class contact between students and faculty.

The faculty are required to have 6 conference hours each week and to do tutoring.

Q5. Advising of mathematics majors; assistance in career selection, career placement and selection of graduate school.

Each student has been assigned one of 8 advisors. The student meets with his/her advisor every semester for course assignment and folder updating. Most students take the minimum 40 hours. In the week before the site visit, the departmental seminar was devoted to career placement. The week after the visit a campus-wide job fest for summer internships and co-op positions was planned. Many students are leaning toward applied mathematics so they can go to work in industry after graduation. Some students go on to graduate study at Southern University or elsewhere. Faculty constantly look for students who can pursue graduate

degrees and encourage them throughout their undergraduate program. It is felt that the transition to graduate school is eased for students who go to graduate school at Southern before pursuing the PhD. The graduate program has 9 full-time (6 from Southern) and 14 part-time (5 from Southern) students.

In October of 1993 Southern University hosted MATHfest III whose purpose was to directly influence the number of underrepresented American minorities who pursue graduate study, leading to earning the doctoral degree in mathematics, as well as to get degree granting institutions to make commitment to provide a mentoring, nurturing, and supportive environment to assist these students. This conference introduced undergraduate mathematics majors (juniors and seniors) who are American minorities to a unique community of professionals in the mathematical sciences: minority American mathematicians who have established exemplary careers, minority American graduate students who are currently doing graduate study in mathematics, and representatives from doctoral granting institutions who are ready to assist, nurture, guide, and support the undergraduate American minority mathematics majors who are willing to meet the challenge of doing graduate work in mathematics.

Q6. Extracurricular activities, e.g., mathematics club.

The department has a Math Club, a Pi Mu Epsilon chapter and an MAA Student Chapter—the Jaguar Tent Club, with 53 members. Members of the student chapter participate in regional and national competitions. Under the sponsorship of one of the faculty, six of these students are doing research and have published solutions to problems posed in the *Mathematical Monthly*. Mathematics majors are involved in the Math Festival contest annually for local middle and high school students.

Q7. Mathematics Help Center and similar walk-in help programs.

The Mathematics Lab is open 8am–8pm M–Th and 8am–5pm F. It is staffed by 2 faculty with release time, 2 full-time employees and 4–5 graduate student tutors, and faculty volunteers. Considerable software exists on the computers which are networked. A lot of testing for students uses home-grown software. Help covers developmental mathematics to calculus and statistics.

Q8. Level of innovation in curriculum and teaching methods.

Curriculum reform is taking place at both the pre-calculus and calculus levels. About half of the calculus sections—4 Calculus I, 4 Calculus II, 2 Calculus III—are using the Harvard curriculum. All of the calculus courses will be taught using this curriculum starting Fall 1996. After the department chair attended the Harvard program, workshops were set up for the faculty. Some faculty volunteered to teach the reformed sections; she asked some. Except for the calculus classes, there is only sporadic group activity.

In the past three years under the Louisiana SSI program, 30 middle school teachers per year have received enhancement in geometry, probability/statistics, and the use of graphing calculators. The department connection with school mathematics is through consulting, the middle school teacher program, science fair judging, and the Mathematics Festival.

Q9. Ways to recruit students to become mathematics majors, through freshman mathematics classes, extracurricular means, recruitment materials.

The department sends letters to local high schools. They then follow-up with prospective majors. The computer science department is separate and has 300 majors, about half are senior division students. The loss of computer science from the mathematics department reduced the number of mathematics majors, although many computer science majors minor in math.

The mathematics faculty and alumni fund scholarships at $500 per semester for each of two students. Dow Chemical gives five scholarships per year. The scholarships are attracting better students to major in mathematics.

Q10. Support programs for women and underrepresented groups.

There are a variety of special programs geared specifically for minority students. One example is "Recruiting Minority Students into Science, Engineering, and Mathematics Graduate Programs" funded by the Fund for Improvement of Postsecondary Education. In order to encourage greater numbers of minority students to enroll in science, engineering and mathematics graduate programs, the University of Iowa established collaborative relationships between University of Iowa science, engineering and mathematics departments and science, engineering and mathematics departments in 7 predominantly minority institutions. Southern University is one of those institutions. The faculty collaborations and the resulting undergraduate research visits will establish a pipeline that increases underrepresented minority students' participation in graduate programs. The project includes faculty curriculum collaborations, undergraduate prep courses, and annual conferences in minority enrollments in science, engineering, and mathematics.

Q11. Use of technology.

Calculators are used in classes and all faculty are networked. Testing for developmental students via computer is an option. The computer is used as part of one section of an experimental Calculus I and one section of pre-calculus. *Derive* is available for use by calculus students in the Mathematics Lab. The honors dorm has computers as do library stations. The department also has secured 200 graphing calculators which are available for students to use.

The 1993 TI project mentioned earlier in this report provided overhead projection units and 100 graphing calculators (40 TI-82's, 40 TI-85's, and 20 HP48G's) for classroom use and for workshops for faculty and students. They have also been used in calculus, linear algebra, and statistics classes. Teachers in the school systems in the community were also trained. This was done through inservice in special graduate classes for these teachers.

Q12. Transfer coordination (for two-year colleges).

Three years from now Southern will no longer be open admissions. A junior college will be established in Baton Rouge and jointly administered by Southern and LSU. Southern University at Shreveport is a junior college and most transfer students come from there.

Q13. Facilities for mathematics majors: departmental computing lab, study room, mathematics library.

Undergraduate mathematics majors have space set aside inside the Mathematics Lab with computers and tables. See Q11 for more about computing equipment. Faculty seem to have computers in their offices, but not all utilize e-mail accounts. The department formerly had a mathematics library but needed the space for graduate student offices.

Q14. Programs for evening and part-time students (esp. at two-year colleges).

One-third of classes are offered at night but no calculus. Most graduate classes are offered at night to accommodate working students.

Q15. Independent study, senior thesis, and summer research opportunities; industrial internships.

There is a seminar for mathematics majors, half of whom are doing independent research projects. The seminar prepares them for doing their Senior Theses. There are summer research opportunities and industrial placements off-campus at AT&T, Bellcore, various national laboratories, Dow Chemical. Southern mathematics majors have participated in the Berkeley Summer Mathematics Institute.

Q16. Honors courses and honors program.

There is a formal Honors College in which mathematics majors may participate. The Honors College curriculum is designed for students to meet the requirements for the honors degree without adding additional courses to their curriculum of area of study. The honors curriculum consists of honors colloquia, independent study, and designated-honors courses in the general curriculum and in the student's major area of study.

Q17. Special-topic and 'research' seminars.

The Department sponsors the Leroy R. Posey Seminar for students and faculty. A committee is responsible for planning and organizing the seminar. The topics for the most recent term ranged from Lie Groups and Lie Algebras and Basic Representation Theory to Co-op, Employment, and Graduate School Opportunities for Mathematics Majors.

Q18. Colloquium series; talks by alumni and industrial managers about uses of math.

The department has a monthly colloquium named after a former department chair but has no money to bring in outsiders. Alumni do return on occasion and speak.

Q19. Formal program for student input to department, such as student member of the department curriculum committee.

The department council was replaced by the Executive Committee but may be reconstituted; it had students and faculty. There is presently no formal mechanism for student input.

Q20. Training programs for junior faculty, particularly non-native speaking faculty.

The department has hired 8 new PhDs in the past 2–3 years. A long-time faculty member is designated to support junior faculty. The process is informal. It involves discussions of the students who come to Southern, a campus tour, information on the background of the school, how to handle certain situations without bothering the department chair. The department does place some emphasis on evaluating the English speaking ability of applicants.

Contact: Professor Lovenia DeConge-Watson.

Report on Site Visit to Spelman College

Irvin Vance and Deb Hughes Hallett
March 1994

Overview

Spelman College is a four-year, historically black college for women. It has a strong tradition of encouraging students to continue to graduate work. With about 100 mathematics majors (including 20 joint mathematics-engineering majors) out of a student body of 1800, the Spelman Mathematics Department has succeeded both in keeping those students who arrived as mathematics majors and also in attracting students who never intended to major in mathematics. Our interviews suggest that the fundamental reasons for this success are not programmatic or curricular, but human.

The department flourishes primarily because of the concern, commitment, and mutual respect of its members—both faculty and students. Over and over again, faculty and administrators sounded slightly perplexed at "where all the students are coming from" and then went on to say how they looked forward to teaching because they missed contact with the students, to complain cheerfully that they were going to have to put on a mathematics and science fair without funding because their students demanded it be continued, and to spend part of a department faculty meeting discussing the experiences of their graduates now in graduate school.

When pushed, the mathematics faculty say that their success may come from their unusually good relations with other science departments. They do indeed have good relations—evidenced by the fact that three chairpersons saw us at 24 hours notice—but this is another by-product of their spirit rather than the cause of it. The students, on the other hand, are not perplexed about why they like Spelman or why they are majoring in mathematics: the small classes (30 is large), the personal attention, the knowledge that the faculty's doors are always open, and the fact that the whole institution believes in them.

The students were equally clear about why they majored in mathematics: many liked mathematics before they got to Spelman and liked what they found there. Others expected to major in something else (not necessarily a science), but a conversation with their calculus teacher convinced them to major in mathematics. Interestingly enough, the faculty resist—even bristle at—the idea that they recruited students, but simply say that they are making sure the students recognize their own potential. This is exactly what they are doing and it has a powerful effect on the students.

Outside of regular courses, the mathematics department places a great deal of emphasis on students doing independent study and research: the senior seminar, which all majors take; participation in summer programs; and attendance at conferences. The college makes enormous efforts to get students ready for professional and academic life (internships, courses at neighboring institutions, GRE review, and mock interviews).

The support that is offered is treasured by the students who keep in touch with the department long after they graduate. The faculty worry that they are doing too much hand-holding, that their courses are not tough enough, and that their students may be overwhelmed by the "real-world" of graduate school. These are reasonable concerns which underlie the delicate balance between support and demands, encouragement and expectations, which this department has achieved.

Answers to the Twenty Questions

Q1. Curriculum and course syllabi geared to the needs of typical students; mixture of pure and applied courses.

Needs and backgrounds of students are determined by placement tests. Two are given, one for nonscience majors, one for mathematics and science majors. The nonscience students are all required to take a quantitative studies course, one of which uses the NSF-funded project CHANCE from Dartmouth.

Mathematics and Science students usually start in precalculus (there are both one and two-semester versions) or in calculus (two varieties, one for biology and economics, one for mathematics and science, are given regularly; there is no honors version). The curriculum in these courses is traditional, but well-thought out. After Calculus II, some science students continue with Calculus III and differential equations. The mathematics students must also take a "foundations" course which introduces them to proofs. This is a relatively new course, instituted to ease the transition to upper level courses and because the department wanted to make sure students learned what "real mathematics" was like early. The course has helped students but has probably not yet hit its stride as a significant number of students must attempt the course more than once.

The upper level courses, Real Analysis I and II and Algebra I and II, have conventional syllabi, similar to those at comparable institutions. The differential equations course, as well as others, have recently added a significant computer component, often *Maple* based.

The courses are largely theory-oriented, but courses in mathematical modeling and numerical methods, plus strong links with other departments and summer programs in industry, keep applied problems within view.

Q2. Teaching styles and active participation of students in class.

Spelman classes are not passive lectures: the classes are small and the faculty and students are used to talking to one another.

High quality teaching plays an important role in promotion and raises—though not the only one. Junior faculty seem to adopt, or adapt to, the culture quickly.

Q3. Requirements of the mathematics major and, if present, secondary mathematics teaching programs.

All liberal arts colleges in the U.S. face a dilemma: how to encourage students to go on in mathematics at the same time as preparing them for graduate school where the average student was educated in countries where specialization started much earlier. Spelman is no exception. However, the department is aware of its dilemma and vigilant in its attempts to maintain the right balance.

There are students at Spelman intending to be high school mathematics teachers. They take the same courses in the mathematics major and meet the other requirements for certification through the Department of Education.

Q4. Nature of out-of-class contact between students and faculty.

Spelman's most impressive feature is probably the outstandingly good relationship between students and faculty. Most impressive, it is not just some faculty and students; it is *all* (or at least all that we met) faculty and students.

Q5. Advising of mathematics majors; assistance in career selection, career placement, and selection of graduate schools.

Advising is another area in which Spelman is outstanding. The faculty obviously take advising seriously as a natural part of taking their students seriously. It is probably no accident that the day we visited a former Spelman student who was about to leave graduate school was back at Spelman for advice as she prepared to go on the job market.

The academic advising is done by faculty (everyone takes part), with career and professional advising shared by other offices on campus. There are the usual health-careers and career planning offices, and in addition an innovative Office for Science, Engineering and Technical Careers (OS-ETC) which specializes in graduate school advising and gathers information on all kinds of internships and special programs.

Q6. Extracurricular activities, e.g., mathematics club.

There are three clubs in operation: a student MAA chapter and a mathematics club (which run together) and Pi Mu Epsilon. Students are encouraged to give presentations at meetings. (Part of the faculty meeting we attended was devoted to a discussion of how to arrange transportation to the next regional MAA meeting and how to arrange student lodging so that those who had funding covered those who didn't.)

There are several locally hosted conferences and fairs including "Math-Fest" at which graduates from historically black colleges return to talk about mathematics based careers and a Mathematics-Science Summit for teams of 5th–12th graders who have "shadowed" Spelman students all year.

Q7. Mathematics Help Center and other walk-in help programs.

In the walk-in Mathematics Laboratory, students can work on the computer or get individual help from mathematics majors. The tutors are selected by the lab director and sign up to tutor specific courses at particular times.

The director regards the center as a way to involve students in the department and hopes that all the department's work-study students will be able to work there in the future.

Some lower courses (precalculus, calculus) have grant-funded tutorials available to students who sign up to come regularly.

Q8. Level of innovation in curriculum and teaching methods.

The CHANCE program in quantitative methods, the use of technology in many courses and the foundations course are innovative. The computer labs in differential equations involve writing; the senior seminar emphasizes writing, speaking and individual projects.

Q9. Ways to recruit students to become mathematics majors, through freshman mathematics courses, extracurricular means, recruitment materials.

See the introduction. There are good written materials but the real recruiting tool is the faculty who are wonderful role models and whom the students respect. Spelman has a summer program for some high school seniors coming to the college which has the effect of introducing incoming students to the department.

Q10. Support programs for women and/or under represented groups.

The support is magnificent. The only criticism might be that there is too much, but it is much appreciated by the students.

Q11. Use of technology.

Over the past couple of years the department has made a concerted effort to introduce technology into as many of its courses as possible. There is a new electronic classroom containing networked computers in which courses can be taught, as well as the mathematics lab which is open for individual work. The department chose to use *Maple* as it was easier to run on cheaper machines, and is now using it in a number of courses. Calculators are being considered for other courses next year.

Q12. Transfer coordination from two- and four-year colleges.

There are about 50 transfer students each year, who are dealt with much the same way as freshmen. They take a placement test, if appropriate, and otherwise are placed individually by counseling.

Q13. Facilities for mathematics majors: departmental computing lab, study room, mathematics library.

There is a departmental computer lab, plus a more sophisticated lab belonging to computer science in the same building. Space for doing anything is limited—one of the drawbacks of Spelman's success is that there is a tremendous space shortage. Not all the faculty have their own offices, some are housed in a temporary building, and nothing is spacious. (There is a new building planned but first funds have to be raised.)

The only topic about which we heard repeated complaints was the library. There is no department library. The college library is shared by the six institutions in the Atlanta University Center (Spelman, Morehouse, Clark Atlanta, Morris Brown, Morehouse Medical School, and the Interdominational Theological Seminary) and is, according to the students, not adequate. The faculty, however, say that the students do not take full advantage of it, especially since the institution of improvements, such as new acquisitions chosen by the faculty and a computerized catalogue.

Q14, 15, 16. Independent study, senior thesis, summer research opportunities, industrial internships; honors courses and programs; special topics and research seminars.

With three categories of independent study courses, there is ample opportunity for independent research during the year and in the senior seminar course. Many students write theses and take part in summer programs. There is a college-wide honors program; honors versions of quantitative studies and calculus are offered, though not regularly.

The spirit of an honors program is conveyed by the federally funded Scholars in Mathematics at Spelman (SIMS) program, which awards partial scholarships for up to 10 mathematics majors who have taken the foundations of mathematics courses and who plan to take a more demanding program than is required by the major (2 semesters each of real analysis and abstract algebra instead of 2 of one and 1 of the other). SIMS students must also take part in an academic summer program.

The new Center for Scientific Applications of Mathematics (CSAM) brings together students and faculty in interdisciplinary research projects.

Q17. Colloquium series; talks by alumni and industrial mathematicians and managers about uses of mathematics.

Talks given in the various Math Fairs, the CHANCE course and CSAM. The faculty is quite free with the phone numbers of former students and often suggest that current students phone them for advice.

Q18. Formal program of student input to department, such as a student member of the department curriculum committee.

There is no formal program for student input into the department but the students' views are clearly taken into account.

Q19. Training programs for junior faculty, particularly for non-native speaking faculty.

There is a college-wide teaching program. New mathematics faculty seem to pick up the culture of the Spelman department very quickly. The department conducts a third-year review for junior faculty, in which their teaching is evaluated.

Q20. Special programs and accommodations for evening and part-time students.

There is a small evening program, for which the department gives a quantitative reasoning course. Regular undergraduate students can arrange to go part time.

Contact: Prof. Theresa Edwards, tedwards@etta.auc.edu

Report on Site Visit to University of Chicago

Susan Forman and David Lutzer
October 1993

I. Introduction to the University of Chicago

To understand the rest of this report, one needs to keep in mind the special mission chosen by the University of Chicago, and the special nature of its students. Chicago must be close to unique in the nation in terms of its history of undergraduate education. Originally conceived as a graduate research institution, Chicago has added undergraduate programs which have grown to national distinction. This contrasts with, for example, Harvard and Yale whose history is just the opposite. Today, Chicago has 9,500 students of whom 3,300 are undergraduates.

Chicago is one of the nation's major research universities in almost every discipline. In mathematics, it is consistently ranked in the top five. It has also one of the lowest student/faculty ratios—about 6 to 1—in the country. There are 50 full-time faculty in mathematics.

It is one of the nation's very selective undergraduate institutions. The average SAT total for the 1000 entering freshmen is 1300. Over 60% of Chicago graduates go on for MS/PhD study, a much higher percentage than the top Ivy League universities. This percentage confirms the mission statements that we heard from the Dean of the Physical Sciences Division, the undergraduate program directors in both Mathematics and Physics, and several Mathematics faculty: departments at Chicago want to produce undergraduate majors who will later become college and university professors in their disciplines.

Out of the roughly 830 students who receive bachelors degrees from Chicago annually, about 40 are mathematics majors per year or about 5% of all graduates (compared to a national average of roughly 1.8%). This percentage is higher than the percentage of mathematics majors at any of the Ivy League universities, according to Project Kaleidoscope data.

Undergraduate mathematics at Chicago at all levels intentionally retains a graduate school emphasis. When asked how they decide whether the department had done a good or bad job with a given graduating class of mathematics majors, the director and associate director of the undergraduate program (Paul Sally and Diane Herrmann) replied without hesitation that the department's judgment is based on the percentage of students who go on to PhD programs, where they go, and what theorems they eventually prove. One of us sat in on a tutorial section run by a mathematics major for five students in the honors Calculus I course. Over and over again, the tutor referred to mathematics graduate school, as in the phrase "Even when you get to graduate school in mathematics, your proofs need to start with . . . ." Of Chicago's mathematics majors in a typical year, about 50% go on to PhD study in group I mathematics departments, and 25% to PhD study in other disciplines. Very few students go on to post baccalaureate employment in industry.

In today's world, the Chicago mathematics curriculum is just as unusual as the university's students and mission. The mathematics major at Chicago is unashamedly a pure mathematics major, straight out of the 1960s. Tracks in Applied Mathematics, Mathematics Education, and Computer Science do exist, but in a typical year, only nine or ten of the roughly forty majors pursue these tracks.

We asked the undergraduate program director what were the most important things from the University of Chicago program which might be exported to other mathematics programs. He identified the departmental commitment to small classes at all levels, the department's program to prepare graduate students to teach undergraduates, and the department's multi-track calculus system and associated successful placement system. The answers to some of our twenty questions will contain details on these three programs, but a preliminary paragraph on each will be helpful in understanding what follows.

Ultra-small classes—five or six enrolled in each section—are the norm at the pre-calculus level. Calculus courses range in size from sections smaller than ten students to "monsters" of size forty. Outside of the honors calculus sequence, all pre-calculus and calculus courses are taught by graduate students. All courses beyond the freshman year are taught by full-time faculty, usually with the assistance of a graduate student, typically in sections of size fifteen or less. The small size is achieved in part by the extensive use of graduate student teaching.

The department is proud of its program for preparing graduate students to teach undergraduates. It is called the College Fellows Program. Here is a typical graduate student career pattern. In the student's first year, there is no obligation to teach. In the second year, all students become College Fellows and are assigned to work with a senior faculty member in a particular course. In the first quarter, College Fellows attend their mentor's class, hold office hours, grade homework, and design and grade parts of the faculty member's exams. In the second quarter, College Fellows are expected to lecture several times in the faculty member's class, in addition to continuing duties from the first quarter. In the third quarter, College Fellows are expected to lecture for approximately two weeks in the faculty mentor's class. In addition to receiving feedback from the faculty mentor, the College Fellow's classes are visited by the department's undergraduate program director who decides whether the graduate student is ready to teach in his/her third year.

Some graduate students are assigned to repeat the College Fellows program, based upon the recommendation of their faculty mentor and the undergraduate program director of the department, but most begin teaching in their third year. Our conversations with graduate students lead us to believe that in most cases the College Fellows program works as it was intended to work.

Placement of entering freshmen into one of three calculus tracks is based upon a three hour departmentally designed test taken by every undergraduate at the University of Chicago during freshman orientation. We reviewed the test and thought of it as a demanding test of traditional single variable calculus. Topics such as working with limits, integration techniques and convergence of power series are covered, and graphing is a major part of the examination. This is a paper and pencil test and the use of graphing calculators is not allowed. The department trusts its placement test and over the years has done at least qualitative studies to verify the test's reliability. The Physics and Chemistry departments also trust the mathematics placement test, using its results for their own placement purposes.

There is an optional three hour test that students who want advanced or honors placement in calculus courses can take. It is a much more theoretical test, and includes both epsilon-delta proofs, applications of the mean value theorem to prove various things, and questions of the type "Prove that $-(a - b) = b - a$." (That last example is not an actual question from the test, but it is true to the spirit of the second placement test.) Based upon the results of the placement test(s), students are advised to enroll in one of three calculus tracks. See Q1, below, for details on the various tracks.

There are two faculty members, Paul Sally and Diane Herrmann, who are giving their lives to undergraduate mathematics at Chicago, and who seem to be the cause of its considerable success. Few of the other fifty mathematics faculty members seem deeply involved with undergraduates. We posed the following question to several people: what would happen were Professors Sally and Herrmann to leave the university?

We heard two answers. The first is that Sally plans to be around for at least fifteen more years and Herrmann for thirty, so what's there to worry about? The second answer involves recalling the department's history. Until the mid-1970s, there were three senior faculty members who were the heart of the undergraduate program, and then they retired. The department needed to find someone else to run the program. Up to that point, Paul Sally had not been centrally involved at the undergraduate level, but he nevertheless agreed to take on the task of directing the undergraduate program. The same thing, it is said, will happen again once Professors Sally and Herrmann leave, since the department deeply values the success of its undergraduate program and will find a way to keep it healthy.

Answers to the Twenty Questions

Q1. Curriculum and course syllabi geared to the needs of typical students, and mixture of theory and applied courses.

Most of Chicago's students enter with adequate pre-college mathematics. As explained above, mathematics placement is based on the scores (and subscores) on departmental placement tests, the first of which is taken by all students. In 1991–92, only fifty students enrolled each quarter in a pre-calculus sequence, to prepare them for the calculus-level two quarter experience that is a graduation requirement. There are three tracks of calculus at Chicago. The lowest is Mathematics 131-132-133, and is typically taken by students who plan to major outside of the physical sciences and/or have a background that is found to be weak in transcendental functions through placement testing. The standard sequence is Mathematics 151-152-153. The Honors Calculus Sequence is Mathematics 161-162-163.

What are the differences between the various calculus sequences? The department gives a sheet titled "A Guide For The Perplexed" to all entering students. Put succinctly, all of the calculus courses involve theory. The 130-level courses require epsilon-delta proofs for linear functions and the 150-level courses do the same for quadratic functions. The 160-level courses are really baby analysis courses, taught from Spivak's *Calculus* with the presumption that students either already know manipulative calculus or can learn it on their own. Students can easily move downwards, e.g., from the 160 to 150 sequence, and sometimes upwards. Most mathematics majors come from the 150 and 160 sequences. Each of these full-year sequences deals only with single variable calculus. Multi-variable calculus appears as part of the second year analysis sequence.

In recent years, the department has begun to offer an alternative to calculus for students who are sure they do not want one of the standard courses. The alternative is Mathematics 110 and 111, called "Studies in Mathematics I, II." To enroll in these alternative courses, students must have placement scores that would allow them to register in one of the calculus sequences. The first alternate course includes a wide range of topics from number theory, geometry, and probability. The second course introduces students to limits, sequences, series, convergence, and a little bit on the derivative. Each of these alternatives involves theory and proofs.

At the sophomore level, there are several analysis courses offered. One is mathematical methods in the sciences, with separate sections offered for physical and chemical scientists. Another deals with mathematical methods in the social sciences. All of these methods courses have a full year of single variable calculus as a prerequisite. For mathematics majors, there are two analysis tracks at the sophomore level. The standard sequence (Mathematics 203-204-205) begins with a quarter on the topology of n-dimensional Euclidean space, which is followed by a quarter dealing with functions of several variables, partial differentiation, multiple integrals, and the implicit and inverse function theorems, and a third quarter focusing on line and surface integrals and the theorems of Green, Gauss, and Stokes.

There is an alternative second year sequence whose prerequisites are "by invitation only" and which was described by the Undergraduate Program Director in the Physics department as being the hardest undergraduate course in any department at Chicago. Topics covered vary with the professor but typically include Banach and Hilbert spaces, measure theory, operator theory, Lebesgue integration, calculus on manifolds, and Fourier analysis. The student-written advising handbook writes "Expect this course to be very intense and to be one of the most challenging courses you will ever take ... You may also be allowed to skip several other mathematics courses after completing [this course]." As was the case in the 100-level calculus courses, students who find themselves out of their depth may easily move from honors analysis to the standard second year analysis sequence.

Once students complete a year of sophomore analysis, they move on to a year sequence in abstract algebra (either in a standard or honors sequence), followed by at least two electives in mathematics. The catalog contains a lot of topology and geometry, some logic, complex variables, theory of computation and combinatorics (the latter two courses being taught by the faculty of the Computer Science graduate program). There is a separate department of Statistics which offers an undergraduate degree and whose courses may be taken by mathematics majors as electives, above and beyond the eight post-calculus courses required for a mathematics major. In addition, a wide spectrum of graduate courses is available to mathematics majors at Chicago, and many begin their graduate studies while still in their first four years. While there are courses in such areas as combinatorics and occasional topics courses in Mathematical Biology, the department's real emphasis is on core mathematics.

The theoretical emphasis in courses designed for mathematics majors causes some problems for other science departments. The Undergraduate Director in Physics described a problem with the sophomore analysis sequence in mathematics: because many of the best physics majors want to hold open the possibility of majoring in mathematics, they want to take the mathematical analysis course instead of the course on mathematical methods in physics. As a result, these students must learn most of the traditional vector calculus on their own.

Q2. Quality of instruction; active participation of students in the learning process.

We reviewed the student-published book of all student course evaluations and found that students see the mathematics department's teaching in a very positive light. They describe the teaching as good, often exciting, and rarely dull. Chicago makes its hiring, tenure, and promotion decisions based upon research performance alone, except (according to the department chair) in a case where someone's teaching is very bad.

Just as the departmental curriculum is very traditional, so are the teaching methods used. Faculty and graduate students use the lecture method in their teaching and believe it is very successful. (So, apparently, do the students.) The department chair expressed exasperation when we raised a question about introducing new teaching methods into Chicago's curriculum—in his view the current system does not need fixing.

However, classes are so small that lectures can easily be conversations. Also all classes in the 130 and 160 series have required tutorials taught in sections of five students. These tutorials are very active, nontraditional learning experiences for students.

Q3. Requirements of the mathematics major and, if present, secondary mathematics teaching programs.

Requirements for the mathematics major are eight quarter courses beyond single variable calculus, including full year sequences in analysis and abstract algebra. Most students design major programs which feature a year of theoretical analysis, a year of abstract algebra, and a year of topology and geometry, and most go beyond departmental requirements and take a wider spectrum of pure mathematics courses, some at the graduate level. Certification as a precollege teacher is possible through a five year program leading to a joint BA/MAT degree. One or two mathematics majors do this each year.

Q4. Nature of out-of-class contact between students and faculty.

Most senior faculty members at Chicago are not deeply involved with the undergraduate program. However, the majority of faculty teach at least one undergraduate course per year and the small size of Chicago classes allows student-faculty relationships to be established through class work. There were few complaints in the student course evaluation book about faculty not being available for out of class office hours. Many seem to have an open door policy, at least on certain days of the week.

There are a few faculty members who seem to devote their lives to the undergraduate program at Chicago and there is a great deal of contact between these faculty members and undergraduates, as noted in the question on advising.

The department holds daily teas for faculty and graduate students; undergraduates are invited on Fridays. At the Wednesday afternoon tea that we attended there were half a dozen faculty members, dozens of graduate students, and several undergraduates who had been invited to meet us there. The Friday afternoon tea is, we were told, attended by dozens of undergraduates, since it is a prelude to the weekly Mathematics Club meeting. The department sees this social hour as an important part of its life.

We would like to generalize Q4 to focus on undergraduates' out-of-class contact with the department and not just with the faculty. This would include contact with graduate students and with other mathematics undergraduates. In this broader area, Chicago seems to do a very good job. Faculty and graduate students hold extensive office hours, run problem sessions for courses and talk about mathematics with undergraduates at tea time. To see the importance of this kind of contact at Chicago, recall that the department is urging its students to think of themselves as pre-graduate students right from the start, and that a remarkable percentage (75%) of the department's majors will pursue PhD study in mathematics or another discipline after graduation.

The more general version of Q4 would include undergraduate mathematics students' contact with each other. One student told us that she did not feel alone as a calculus student at Chicago because everyone at the university seemed to be taking calculus. Other student-to-student contact among mathematics majors is fostered by a departmental program that pays undergraduates to run help sessions for other undergraduates, and by the department's decision to use its best undergraduates to run the five-student tutorial sessions in its 130 and 160 level calculus courses. This, too, is consistent with the department's goals for its students— to induce them to become teaching mathematicians. The same comment applies to the department's decision to hire its best undergraduates to work in the Younger Scholars outreach program. (See Q9 for details.)

The net effect of all this involvement is to tie undergraduates into the department, almost as apprentice colleagues, and to bring them into closer contact with each other. This creates an intellectual community and that sense of community is further strengthened by the existence of study groups in mathematics, especially at the junior and senior level. One student commented that having all mathematics courses in the same building was also a community-building factor since it caused students' paths to cross frequently.

Normally one would expect the Mathematics Club and the annual Putnam team to be additional opportunities for student-faculty interaction outside of class. This is not the case at Chicago since both the Mathematics Club and the Putnam team are entirely student run activities, with faculty members not invited.

Q5. Advising of mathematics majors; assistance in career selection, career placement, and selection of graduate schools.

Throughout their undergraduate careers at Chicago, all students have official advisors in a professional advising center run by the university. In addition, the department has two faculty members who are deeply involved with advising, Paul Sally and Diane Herrmann. These two run the placement system and advise students about upper division courses and about graduate school placement. They seem to know the mathematics majors by name, and the students know them.

Career advising is problematic for those student not interested in doctoral study. However, graduate school advising appears to be very good. The department measures itself, to a degree, in terms of which Group I schools their majors get into. Professors Sally and Herrmann are largely responsible for graduate school advising and attempt to match a given student's abilities and interests with the admissions standards and scholarly strengths of different graduate schools. In addition, the department arranges annual open meetings for juniors and seniors in which faculty members from other University of Chicago departments make presentations about what a mathematics major can study by going to graduate school in the other discipline. This is a reasonable response to, and perhaps a contributor to, the fact that 25% of the department's majors pursue PhD study in another discipline.

Q6. Extracurricular activities, e.g., mathematics club.

The department has a Mathematics Club and a Putnam Team. Both are entirely student run. In the Mathematics Club, which was revived several years ago by the students after a several year hiatus, students give weekly thirty minute talks on mathematical topics that they have learned outside of class. In a recent year, student talks have included topics such as "The Moebius Function of a Finite Group" and "On Counting Labeled Trees in Your Sleep: Two Astonishing Proofs." There is no COMAP team and little interest in starting one.

Q7. Mathematics Help Center and other walk-in help programs.

The Mathematics department has an extensive program of problem sessions and help sessions which it runs in its own building, Eckhart Hall. This is one way for undergraduates to earn money from the department for doing mathematical things, and to interact with each other. In addition, most dormitories also have mathematics help sessions, paid for by some other office. Finally, the duties of second year graduate students (i.e., the members of the College Fellows Program, above) include holding office hours and running problem sessions for students enrolled in the courses of the College Fellows' faculty mentors. We believe that there is a more than adequate program of out-of-class help available to students.

Q8. Level of innovation in curriculum and teaching methods.

As noted in the Introduction, the University of Chicago mathematics curriculum is a model of what the earliest CUPM reports had in mind, and it seems to work very well. Chicago admits that it has not adopted calculus reform. On the other hand, writing has been an important part of the mathematics curriculum at Chicago for many years. For example, the calculus placement test asks students to write their answers in English sentences, and we were told that writing is important in all calculus courses at Chicago. When one of us visited a tutorial for the honors calculus, Math 161, the undergraduate tutor was insisting that students at the blackboard write in full English sentences, rather than simply put down formulae.

Q9. Ways to recruit students to become mathematics majors, through freshman mathematics courses, extracurricular means, recruitment materials.

The Mathematics department is deeply involved in high school outreach activities in the Chicago area public schools. These include a program called the "Younger Scholars Program" (YSP) which brings about 100 exceptionally talented junior and senior high school students to the University each summer, with twelve follow up sessions on Saturdays during the academic year.

In addition to working with gifted pre-college students, Professors Sally and Herrmann also stay in contact with the Metropolitan Mathematics Club, an organization of high school teachers from the greater Chicago area. Professor Robert Fefferman, another department member, also runs a summer program in which high school teachers bring their best high school students to a six week long summer mathematics program at the University of Chicago. Keeping the University's name before these teachers through these various programs leads to a continuing stream of strong students from local high schools.

The department publishes a four page flyer about majoring in mathematics at Chicago. Once a student reaches campus, there is some mathematics major recruiting material provided by the department. A student group publishes a book of course evaluations annually, and the reviews of mathematics courses are positive enough to serve as recruiting materials themselves.

Q10. Support programs for women and/or under represented groups.

The mathematics department provides outstanding support programs for all of its students. There are not any special programs for women or minorities. The women students with whom we spoke did not have any complaints about the atmosphere in the department from the faculty, but did mention occasional insensitivities on the part of male undergraduates. Like most elite institutions, Chicago's percentage of female mathematics majors is around 25%, rather than the higher national average of around 45%.

Q11. Use of technology.

The University of Chicago is not a leader in the use of technology in mathematics or mathematics teaching. In courses such as Numerical Analysis, computers are used and the department sees a role for computers in differential equations courses. But in most other courses, the faculty believe that their theoretical emphasis does not lend itself to computer use. Chicago undergraduates have no qualms about using *Mathematica* to solve homework problems, and faculty members reported several innovative student solutions of theoretical homework problems using *Mathematica,* something to which the faculty members do not object. When pressed, several argued that technology should be used if older methods of teaching are not working, and quickly added that at Chicago, the older methods still work.

Q12. Transfer coordination from two- and four-year colleges.

Few transfer students.

Q13. Facilities for mathematics majors: departmental computing lab, study room, mathematics library.

The department does not have, and does not seem to want, its own computing laboratory (see Q11, above). Mathematics majors have access to computer laboratories around campus, and that seems to be adequate.

The department has its own research library which is a powerful mathematical resource and which is used by undergraduates both as a study room and as a journal reading room. Chicago seemed somewhat unusual to us in this regard. Normally, most undergraduates in mathematics use only textbooks. Several of the undergraduates with whom we spoke at Chicago described taking courses based on current journal articles, without textbooks.

Finally, the department has several meeting facilities for its students. One is called "The Barn" and the other is an impressive complex of rooms that is used for the department teas that were mentioned in response to question Q4.

Q14. Independent study, summer research opportunities, industrial internships; honors courses

The department offers a course "Proseminar in Mathematics" which is open to mathematics majors, as is the option to write an honors thesis. However, few of the department's majors write theses, preferring instead to begin taking graduate courses during their undergraduate years.

Chicago sends many of its students elsewhere for REU-type summer research opportunities in mathematics. Some summer programs are available for mathematics majors, e.g., internships at Argonne National Laboratories. The department's Undergraduate Program Director commented that a surprising number of his majors participate in summer research experiences in other disciplines e.g., by working in scientific laboratories. Perhaps this partially explains the large percentage of Chicago's mathematics majors (25%) who pursue PhD study in a discipline other than mathematics. Finally, many of the best undergraduates are invited to remain at Chicago during the summers to participate in the department's Younger Scholars outreach program. (See Q9.) In some sense, given the department's goal of producing college professors, this kind of teaching involvement might serve the same function as do industrial internships elsewhere.

Q15. Honors courses and programs

The freshman honors calculus course attracts a large number of students, 125 to 175 a year. This may be due in part to the high standards for placing out of first-year calculus. As noted above, the syllabus for standard sophomore analysis course would make the course an honors course at virtually any other university. There are upper division honors courses and an honors program, but the better students prefer to take graduate courses instead.

Q16. Special topic and research seminars.

Samples of special topics courses are: in the spring quarter of 1993, Introduction to Mathematical Biology; in the autumn quarter of 1993, Multilinear Algebra and Mathematical Methods and Models in Population Biology. Mathematics majors at Chicago also see the department's graduate program as a list of special topics courses that are open to them.

Q17. Colloquium series; talks by alumni and industrial mathematicians and managers about uses of mathematics.

There is an annual Paul R. Cohen lecture on a mathematical topic accessible to students in the Math 160 honors calculus sequence, but we saw no evidence of talks about post-baccalaurate employment opportunities.

Q18. Formal program of student input to department, such as a student member of the department curriculum committee.

There is no student membership on departmental committees. At the level of the Dean of the Physical Sciences Division, there is a student advisory committee which, among other things, prepares advising materials about the various physical science disciplines, including mathematics. The forty-page pamphlet written, published, and distributed by this student group is very good.

Q19. Training programs for junior faculty, particularly for non-native speaking faculty.

The Mathematics department invites its new faculty to participate in an orientation session at the beginning of the year. These sessions are required for second year graduate students, all of whom belong to the department's College Fellows Program that is described in the Introduction to this report. Some of the younger faculty attend, but many do not.

The undergraduate program director gives special attention to graduate students and new faculty who are not native English speakers. We heard one report of a stringent training process which requires non-native speakers to give an undergraduate mathematical lecture to the Undergraduate Program Director who records the lecture, critiques it harshly, and then sends the lecturer back to prepare another, hopefully better, lecture. In addition, the Associate Director of Undergraduate Programs visits each 100 level class each year, to monitor teaching quality.

Q20. Special programs and accommodations for evening and part-time students.

None

Contact: Prof. Paul Sally, sally@zaphod.uchicago.edu

Report on Site Visit to University of Michigan

Harvey Keynes and William Lucas
April, 1994

1 Lower Division Program

Most of the exciting changes that have already been implemented at Michigan are occurring at the lower division level. This includes major changes with their "new wave" calculus, interesting experiments in their several different honors courses and an alternate to calculus, as well as revisions in pre-calculus and the course for those not planning on taking a calculus sequence. As a large university with a great number of calculus sections each semester, Michigan has traditionally had several different tracks for completing the first two years of mathematics. Most of these course sequences are currently involved with new experimentation.

A. Math 115. The greatest change is taking place in the first year of the three courses in the standard analytic geometry and calculus sequence. A few years ago, this sequence began using graphic calculators (Tl-81's), assisted by a $400,000 NSF grant directed by Morton Brown. The grant provides training for faculty on how best to use the calculators. Michigan currently uses the Harvard text for the first two semesters of this sequence.

Some 20 of the many sections of Math 115 are taught in small groups, are highly interactive, and use a cooperative learning format. The classrooms have been redesigned for 24 students with four around a table who work together throughout most of the class. They work on problems and have a weekly group project. The groups change every few weeks. (The 24 per class is down from the former class size in the low 30's.) The courses stress problem solving and quantitative understanding via collaborative learning. Geometrical insight is emphasized. This experiment has been well documented and assessed somewhat independently via the University's Center for Learning and Teaching. Initial evaluations indicated some success, and students appear pleased with it and surely feel their teachers are more sincerely interested in their learning. Teachers of subsequent courses have not noticed any decline in abilities of students coming out of these experimental sections. More tradi-

tional sections also exist and they too are using the Harvard text.

The small-group "new wave" approach was initiated by Professor Mort Brown. Senior lecturer Patricia Shure led the team that developed the training materials for instructors. The current efforts are being directed by Brown and Shure. Ample help opportunities outside of class are also available for this and other courses, mainly via the Math Lab (discussed below). [*Update Note*: In fall 1995, the "new wave" approach will be implemented in all sections of Math 115, 116.]

B. Math 116. Math 116, which is the sequel to Math 115, is taught in the same style as Math 115. However, in spring 1994 seven sections of Math 116 were experimental sections which have been going on for a few years. These special sections directed by Professor R. Wasserman involve hands-on modeling, open-ended projects, and working in groups of four students that change every couple of weeks. The students have a lab using MAPLE as a tool for problem solving. The classes are about 60% engineers and 40% others, and about two-thirds of the students do reasonably well. In the fall this course has many students with AP placement who are often bored in the traditional 116 and not ready for Math 215, the multivariable third semester course. It appears as a reasonable solution for this rather common difficulty at most schools.

C. Math 215. Math 215, the sequel to Math 116, has larger class sizes (80 students) with three lecture and one recitation periods. This year they have introduced some newer experimental sections that make use of MAPLE, but again primarily as a tool. There is a lab manual for students and some summer training for faculty and TA's. An excellent set of notes on Computer Laboratory Projects for Multivariable Calculus by E. A. Gavosto and A. Uribe have been developed for this interesting course experiment.

D. Elementary Courses. There have also been major revisions in the standard pre-calculus course (Math 105) which includes use of calculators, cooperative learning, modeling, help at the Math Lab and gateway exams. They have also modernized their sequence for non-science majors: Math 127 (Geometry and the Imagination) and Math 128 (Explorations in Number Theory). This appears to be an outstanding attempt to deal with a pending numeracy requirement by means of an interesting course combining reasoning and culture. Math 127 is based on similar courses at Brown and Princeton (the more recent Thurston course rather than previous A. W. Tucker one). It uses Banchoff's book *Beyond the Fourth Dimension*. The material shown the site visitors was imaginative and carefully developed. Student reaction was very positive, and continued development of these courses is ongoing. These are excellent courses for preparing elementary teachers.

E. Honors Sequences. There are two (four semester) honors sequences for analytic geometry and calculus: Math 185-186-285-286, and the more intensive, theoretical, abstract, and rigorous 195-196-295-296. The latter sequence has recently used books by Spivak, Apostol's Volume 1, and Courant and John. Many of these students are in the College of Literature, Science, and the Arts' (LSA) honors program. Mathematics majors are often recommended to take these (or the following alternative). Many of the students from these honors courses do go on to graduate school, but not particularly many in mathematics programs.

F. Math 175-176. This is a rather new special sequence in the mode of "alternatives to calculus" which are being tried at some schools. The goal again is to get students to think for themselves, build their own models, and gain confidence in problem solving. Students often arrive at the answer first and then question the reasons or concepts. Computers are used freely. About two-thirds of the students are in the honors program and some calculus from high school is presumed. The students' note books are collected every ten days.

Most of differential calculus is covered, but some aspects of integral calculus are left out over the two semesters. The main new ingredient is a major dose of combinatorics (in a fairly classical sense) and a bit of probability, including enumeration, recurrence, graphs and codes, and asymptotics and chaos. Some finite differences and dynamical systems are covered. Devaney's book along with ample

notes by Phil Hanlon (and C. Greene, Haverford College and Joan Hutchinson, Macalester College) are used. The visiting team felt that this course would be an even more attractive alternative if the course established a better identity by being able to more clearly articulate what students should know at the end of the year. The contents of this course was the topic of a conference entitled Discovery and Experimentation in the Freshman Mathematics Curriculum held on June 6 and 7, 1994.

Faculty are currently discussing the type of second year course that should follow 175-176. They are considering emphasis on a unified approach to real world applications, although many aspects of linear algebra and some optimization would likely be included.

Phil Hanlon's exceptional efforts illustrate what several distinguished research professors are currently doing at Michigan to enhance undergraduate teaching.

G. Math Lab. They have put substantial resources into a "Math Lab" which is primarily a tutoring area for freshmen and sophomore courses (through linear algebra and differential equations). With several thousand students, this is necessarily a large operation which involves some 25 paid undergraduate assistants, along with all TA's and some junior faculty volunteering an hour (or more) per week. It is open 39 hours per week with six or more helpers available at all times.

2 Upper Division Program

As a major university, Michigan has a large and rich offering of courses in mathematics: some 55 different upper division courses per year plus about 25 graduate courses which some advanced undergraduate students take. In addition to the typical pure mathematics major, plus the honors program, they have developed a mathematical sciences major. They also have an actuarial program and provide for a teaching certification.

A. Mathematical Sciences. This program aims to provide a broad mathematical training plus specialization in some applied direction. The six areas of concentration are:

(i) Discrete and Algorithmic Methods;

(ii) Numerical and Applied Analysis;

(iii) Operations Research and Modeling;

(iv) Probabilistic Methods;

(v) Mathematical Economics;

(vi) Control Systems; and

(vii) Mathematics of Finance and Risk Management.

Details of each specialization are spelled out in the Undergraduates Programs booklet for Mathematics (37 pages). They reviewed programs at schools like the University of Washington, Stanford, and UCLA before arriving at their program. There are the usual debates within the department about this degree. Does it have sufficient depth? Does it really prepare students adequately who may go to graduate school in the mathematics sciences? Is it designed for the weaker students? Should every mathematics student be required to take a serious advanced calculus course?

One increasingly popular course is Mathematics 462: Mathematical Models, initiated by Carl Simon. It now meets each semester, has other teachers as well, and the theme varies.

B. Actuarial Mathematics. Michigan has historically had one of the leading programs in actuarial mathematics and the only major program among the dozen top-ranked mathematics departments in the country. This is a very popular program among the students and the department is currently attempting to revitalize it. Three professors in mathematics are involved in this program as well as some from other departments. Mathematics is trying to take advantage of additional resources in fields like public policy and public health, and are considering a dual degree with Business. They are seeking new faculty positions in this direction and are about one-third of the way to a $1,500,000 endowment for a Cecil J. Nesbitt chair in this field. (Although Nesbitt has been retired for many years he is routinely working in his office and still directs REU projects.)

The students take one or two courses in each of accounting, economics, and computer science. There is a core of five basic mathematics courses: differential equations, analysis, probability, statistics, and numerical analysis. There are three special actuarial courses (interest and insurance, plus a year of life contingencies) and two additional courses related to actuarial science (not necessarily in mathematics). It is not difficult for a student to satisfy these requirements and to be a mathematics major.

C. Mathematics Education. Michigan formerly had a strong commitment to education within Mathematics, but this area lost significant resources over the past dozen years when teaching positions became scarce. Two retiring professors were not replaced, leaving only one such specialist. There are now many fewer teachers attending summer school, and the University provides less in terms of service to the regional school districts. Michigan has a large School of Education, and secondary mathematics education students are split between Education and LSA. Only a few of the mathematics majors each year concentrate in this direction. (This contrasts with Michigan State University where about half of the mathematics majors pursue this option.) On the other hand, the Department is trying very hard to improve the situation in education. However, they are not obtaining the needed support from the College's administration. They do seem to be well positioned when the resources do become available from their administration or some national source like NSF.

There are about 90 students a year who obtain a credential in elementary education (K–8) and about 25 in secondary mathematics education (9–12). The former take two (some three) mathematics courses, and the latter take two: Concepts Basic to Secondary Mathematics, and Topics in Geometry for Teachers. Many activities at the lower division level such as the new wave calculus or Mathematics 127 and 128 have much to offer to prospective teachers. Professor Krause who leads the education efforts in the department is doing an outstanding job and is very popular with the students. He clearly needs some help from others.

3 Advising and the Undergraduate Office

In recent years the Mathematics Department has put major efforts into building a complete support structure for its majors. This is an important development, because such assistance that comes rather naturally at most small colleges is often not done well at large universities with over 200 majors. First, it should be noted that several of the leading senior research faculty have expressed the desire and need for improving the undergraduate program, and are devoting substantial time to this effort. There is an undergraduate program committee responsible for changes in the program. There are about seven faculty who serve as the primary counselors for the general majors. They are each scheduled for a couple of hours per week, as well as being available at other times and hold scheduled appointments. (The actuarial faculty, however, usually advise their own majors. In addition Professor Burns is involved with the honors students, and Professor Eugene Krause advises those interested in education.) There is also a 37 page booklet produced annually detailing all aspects about the undergraduate mathematics major.

One very exciting development is the Undergraduate Program Office for the mathematics students (housed in the Cecil J. Nesbitt room). This was set up about five years ago with the financial assistance of an endowment of some $100,000 from alumni. It is staffed full time by a person called a Student Services Assistant. They are fortunate that the current person in this position had prior experience as a general advisor in the College. She thus has expertise in the overall College requirements as well as working on the needs of mathematics majors, and also has accessibility to more general advising data and records. She was not a mathematics major and does not have the role of designing individual programs for the majors. She schedules, coordinates, and monitors the advising which in turn is done by faculty. She also provides assistance to particular students on a variety of other matters.

This Undergraduate Program Office provides additional services and information. It has built a file on career options, financial aid, and graduate programs in mathematics. It is compiling an alumni resource network. It organizes talks by nonmathematics faculty on various areas of concentration for undergraduates. It has assisted in recruiting majors. This includes activities such as a "mathematics fair" and informing students of the various honors courses. The Undergraduate Program Office also serves as a meeting place for students. On the other hand, there does not seem to be much going on in the nature of "math-club-type" activities, except for the fairly well organized actuarial students. This office consists of three separate rooms: a common meeting area, a small computer lab with a few workstations (available only to the majors and used, for example, on REU projects), and an office area with a small advising cubical. They put out an undergraduate newsletter four times per year. We feel that this undergraduate Program Office could serve as a good model for many large university mathematics departments.

4 Research Experiences for Undergraduates

The department has a large and popular REU program for its own students. Students work individually on concrete research problems under the close supervision of a faculty member. About 20 students have been involved in each of the summers of 1991, 1992, and 1993. Additional students have also had such experience off campus. Professor Daniel Burns has done much of the organization which involves recruiting and disseminating information on available faculty and problem areas. It should be noted that Michigan is not a REU "site." NSF has provided substantial funding, but via supplements to individual PIs' research grants. The faculty mentors then may be someone other than the particular PI. Only a couple of faculty direct more than one student per summer. The university has also provided some additional financial support for these activities. The research problems have included topics from actuarial mathematics, business, economics, the auto industry, as well as more typical mathematics. Some students have made presentations at conferences. About 30% of the participants are women. The visiting team was shown a few of the REU project reports, and these were of good quality.

5 Student Activities

The visiting team interviewed two separate groups of about 8 to 10 students: students in the actuarial program and a group of general upper division mathematics majors.

A. Actuarial Students. These students were very cohesive as a group, and quite positive on their involvement in the Actuarial Program. Their linkage to the other mathematics majors and department in general, however, was not very pronounced. They were, as a group, unsympathetic to the new wave calculus, because they feel that it would lack the skill preparation necessary for the actuarial exams and their overall program. They felt that the mathematics sciences degree was generally oriented towards engineering rather than towards their interest in business. Program planning and advice was inconsistent, and they desired more experiences in summer jobs. While the basic link through their actuarial interests keep them together as a group, it would appear that the Department could put more effort into their advising and overall support.

B. General Majors. The same concerns seemed to be echoed by the general mathematics majors. While more sympathetic to the goals of the new wave calculus, they again had concerns about the unevenness of advising, the general level of teaching, and the lack of cohesiveness among the mathematics majors in the department. Several hoped that the Mathematics Club would be more active to help address these issues. Others looked forward to more cooperative learning and group activities in upper division coursework. The Undergraduate Office provided some additional support, but obtained mixed reviews. They supported more interaction with faculty. The Residential College was popular.

Although the number of mathematics majors has been increasing in recent years, the department desires more and is putting serious effort into this goal. Few mathematics majors go on directly to graduate school. One figure we heard was 12 out of 100 last year: 6 in mathematics and 6 to other fields. For 1991–92 it was 14 out of 105: 7(or 8) in mathematics and 7 (or 6) other. They seem to have a rather typical number of women students for mathematics, and a few minority students. Pat Shure interviewed female students and they did not feel they were encountering any specifically gender problems. There is also a "bridge" program for minorities.

The students have done well on the Putnam exam in recent years. We did not hear whether they entered the COMAP Mathematics Modeling competition.

6 Faculty Attitudes and Leadership.

A. Senior Faculty. The site team met with several senior faculty on the first morning and had many other opportunities to discuss the overall climate for participation in educational and curriculum development at Michigan. It is very clear that a real culture shift has taken place in the mathematics department, and that many senior research faculty viewed involvement in undergraduate education as a genuine professional responsibility to be seriously addressed. There are many examples to illustrate this culture shift. Senior faculty in the new wave calculus courses teach four hours per week, plus do their own grading for 5 to 6 hours/week. Such an assignment counts for one semester course in a standard two courses per semester teaching load. Faculty were very confident about their teaching, and eager to discuss involvement in curriculum development. The widespread acceptance and support for different types

of faculty positions, such as a senior lecturer whose primary responsibility is the Calculus program and a tenured faculty member whose primary responsibility is the Math Lab, are strong indicators of real change. The genuine interest in developing a more modern and technologically supported curriculum to help attract more majors was striking, as well as the recognition of the significant funding given to the Department to support their calculus reform efforts. The willingness to become more involved with REU's to attract more majors and keep Michigan competitive for its out-of-state students also indicated a different faculty attitude.

B. Junior Faculty and Post Docs. A separate meeting with junior faculty and research instructors (and several informal conversations) confirmed a similar attitude as the senior faculty. These faculty were clearly told that teaching and curriculum development were very important. Most were confident of their ability to do good research and also put real effort into teaching. They were mentored in both their research and teaching.

Teaching loads for junior faculty were quite heavy for a research university. An assistant professor could teach 8 hours per semester and, if his or her assignment included a new wave calculus course, grade 5 hours per week in addition. Overall, however, the junior faculty saw the value of the conceptual approach and new pedagogy in the new wave calculus, and were willing to put the extra effort into these courses. Some did question whether these reform efforts would produce more mathematics majors or are really worth the effort. But in general, the young faculty were addressing education more seriously and putting more effort into it than at many other research-oriented universities.

C. Leadership and the role of Donald Lewis. It was extremely clear to the visiting team that the major catalyst for the change in culture at the Michigan Mathematics Department was its chair Don Lewis and his extraordinary leadership. In addition to working overtime with his faculty to accept a change in culture towards education, he has secured large amounts of institutional support and hired faculty with different roles to make the environment at Michigan very receptive to educational innovations. Most of the curriculum initiatives were carefully crafted by Lewis and selected faculty. Many are in their early stages of development and, even with faculty enthusiasms for curriculum change, will need continued departmental support to mature and eventually be institutionalized.

There are many issues which are just beginning to be addressed or are not yet even started. These include the desire for more majors and students going to graduate school, the content of the Mathematics Sciences degree and Actuarial specialization, better linkages with K–12 teachers, and more diversity in the student population. Making progress in these areas will require equally dynamic leadership as Lewis has offered over his past two terms. It will be interesting to see if the culture change is maintained as well as the momentum for continued reform, and if the curriculum innovations are ultimately institutionalized under the new leadership in mathematics at Michigan when Lewis retires.

7 Administrative Support

Although the University of Michigan is a large state university with an outstanding research reputation, it has also traditionally been known for its superb undergraduate education. Nearly 40% of its undergraduates come from out of state and pay tuition in the range of well-known private colleges. The students in the College of Literature, Science, and the Arts (LSA) come mostly from the top quarter of their high school class and have a median SAT score of 1100. When the new dean of the College took charge five years ago, she stressed the need to improve undergraduate teaching, and she really meant it! (We met with the College's associate dean for education, M. Martin.) The chemistry department was the first to make major innovations in their beginning courses with a heavily lab based program; and mostly with outside funding obtained by the department. One result has been a large increase in the number of chemistry majors.

For various reasons the administration judged the mathematics program as ripe for change, and has provided a significant amount of the college's financial resources to support it. The college's motives were much more in terms of better education in general and a more interesting student experience (especially at the lower division level), than they were for calculus reform or more mathematics majors per se. This support included reducing some class sizes from the low 30's to 24, the reconfiguring of class rooms for the Mathematics 115 courses, and the hiring of several additional faculty for three-year (non-tenure track) positions. The administration views this as a long-term commitment (ten years) and considers it as successful so far. The move toward smaller classes in mathematics was also consistent with a college wide initiative to have some 200 freshmen

seminars of 18 to 25 students each. In general the Dean of LSA has been extremely supportive of improvements at the freshman and sophomore level, and Mathematics was fortunate to be a major recipient of such aid.

Contact Person: Professor Mort Brown, mort.brown@math.lsa.umich.edu.

Report on Site Visit to University of New Hampshire

Alan Tucker
April, 1994

The focus of this visit was the preservice teacher preparation program for secondary school mathematics teachers. The visit began with a breakfast with Otis Sproul, Dean of Engineering and Physical Sciences, and Ken Appel, Mathematics chair (Tucker had dined and talked at length the night before with Joan Ferrini-Mundy). The Dean was most supportive of Mathematics and in turn is highly regarded by mathematics faculty. Mathematics's location in a College of Engineering and Physical Sciences seemed to be viewed currently as a positive situation, although in the past some said that Mathematics had felt like a second-class service department. The chair is an eminent mathematician from the University of Illinois (co-solver of the famous Four Color Conjecture) who came to UNH last fall on a term appointment of three years. (This appointment is an outgrowth of recommendations of an external review a year ago.)

Following breakfast, the day was filled with a series of meetings with individuals and groups from the mathematics education full-time faculty, mathematics education part-time and visiting faculty, and the rest of the mathematics faculty. There was also an 1 1/2 hour meeting with students. A laboratory session of a linear algebra class and a recitation section of second-semester calculus were attended. There was a final exit meeting with Professors Appel and Ferrini-Mundy.

Twenty-five years ago the 23-member department consisted almost exclusively of pure mathematicians. Now, while the same size, the department has a mixture of pure mathematics, applied mathematics, statistics, and mathematics education faculty. Mathematics education has a more substantial role at UNH than at virtually any other doctoral mathematics program in the country. Perhaps reflecting its mainstream role in the department, doctoral students in mathematics education are required to take the same set of PhD comprehensive exams as all other mathematics doctoral students. Likewise, secondary mathematics education majors take most of the same mathematics courses that regular mathematics majors take. The UNH mathematics education PhD's, while not large in number, have all obtained excellent jobs and some have become very well-known, e.g., Joan Ferrini-Mundy and Tom Dick (Oregon State).

Generally, there seemed to be mutual respect and goodwill among the four subgroups in the department. The groups all have several productive researchers. There is a sense that the values of pure mathematics, or core mathematics as it is called at UNH, have a pre-eminent influence in the department. However, mathematics education also seemed to have a substantial role in the department. Most of the federal funding currently comes from mathematics education and curricularly based ILI-type grants (creative pedagogical approaches from mathematics education faculty appear to have played a role in getting one ILI proposal funded).

There is a large summer Masters of Science for Teachers program that offers summer employment to many faculty in other groups. Through teaching in the MST program and trying to relate their mathematical interests to the needs to the teachers in these summer classes, several mathematics faculty have mathematics education concerns on the minds in their regular academic-year teaching.

The department offers a wide variety of options for a major in mathematics. There are BA and BS degree programs in mathematics; elementary school, middle school, and secondary school BS in mathematics education programs; plus a spectrum of interdisciplinary options, with chemistry, computer science, economics, electrical engineering, fluid dynamics, mechanics, physics, statistics and thermodynamics. The department graduates about 60 majors a year, above average for a doctoral institution awarding 2400 Bachelor s degrees annually.

While the focus of the visit was on the preparation for prospective teachers, it seemed clear that the strength in mathematics education is dependent in part on the department's diversity of mathematics groups and on the mutual respect and interactions among these groups. The engineering college, with its professional orientation, also seems to have a positive influence.

School mathematics teachers are said to copy the teaching styles of faculty who taught them mathematics in college, irrespective of what they are told in their methods courses. At UNH, the preservice mathematics teachers see innovative educational practices in many courses. The mathematics education faculty regularly teach geometry and a senior seminar as well as the methods course and a variety of core mathematics courses. Geometry and the senior seminar are taught almost exclusively through group work. Calculus is taught in large 200-student lectures, but with small recitation sections in which group learning and exploratory problem-solving are emphasized.

As a result of teaching summer courses for teachers and the presence of mathematics educators (and occasional attendance at mathematics education seminars), some of the non-mathematics education faculty seem more open to using innovative curriculum and pedagogy approaches in their courses. For example, most of the first half of a recent abstract algebra course was devoted to examining the algebraic 'wildlife' that exists throughout mathematics. Some linear algebra and differential equations classes do a lot of exploratory learning with mathematical software. This visitor was very impressed at the working strategies he saw in a linear algebra laboratory session that integrated many ways of doing mathematics: students were scribbling sample calculations one minute, then doing visualization and number crunching on workstations the next minute, then maybe turning to a calculator for supporting low-level computation, and occasionally gathering in a group to confer with others for several minutes.

There are two types of non-tenured track faculty used in the mathematics education program at UNH. First, retired mathematics teachers and state Department of Education experts are employed to teach and oversee placement and supervision of student teaching interns. Second, using Eisenhower grant money, there has been a program for high school mathematics teachers to spend a year at UNH doing some college teaching and sitting in on courses and seminars. Both of these efforts build goodwill with local schools, which leads to better placement of student interns and to school mathematics teachers recommending UNH to their better mathematics students.

The mathematics education students at UNH were impressive in their broad awareness and thoughtfulness about issues in mathematics education. They were thinking far above the level of 'how do I make up a good lesson plan.' They knew the NCTM *Standards* document inside-out (it is the 'text' for their senior seminar) and had given a lot of

thought to the pros and cons of different teaching styles.

Answers to the Twenty Questions

Q1. Curriculum and course syllabi geared to the needs of typical students; mixture of pure and applied courses.

The visit was focused on preservice teacher preparation, but in all courses the level and objectives of the courses seemed most appropriate and course syllabi matched these objectives. The curriculum in mathematics and in the numerous interdisciplinary options in the major gave students an excellent spectrum of educational opportunities.

Q2. Teaching styles and active participation of students in class.

These issues are largely addressed above. There was active participation in a variety of modes. Students asking lots of questions, working together on open-ended mathematics problems in recitations and in computer labs, and class discussion to develop new topics in place of classroom lecturing.

Q3. Requirements of the mathematics major and, if present, secondary mathematics teaching programs.

The major requirements were quite prescriptive. The BS mathematics major requires courses calculus I,II, multivariable calculus, differential equations, linear algebra, mathematical foundations, analysis, abstract algebra, computer programming, complex analysis, topology plus two electives The BS in high school mathematics teaching is very similar to the mathematics BS with complex analysis and topology replaced by geometry, a senior seminar and a mathematical methods course. In addition, four education courses and a teaching practicum are required. The requirements for middle school and elementary school mathematics teaching are somewhat lighter mathematically. The preservice mathematics teacher programs were well matched to a thorough preparation to teach to the NCTM *Standards*.

Q4. Nature of out-of-class contact between students and faculty.

There was so much student-faculty interaction in classes that when students spoke about discussions with faculty I did not know whether it was in-class or out-of-class.

Q5. Advising of mathematics majors; assistance in career selection, career placement, and selection of graduate schools.

Almost all students I met were preservice teachers and they all had good advising. One would expect that career advising was good for all students.

Q6. Extracurricular activities, e.g., mathematics club.

There is a Pi Mu Epsilon Chapter. Not too active.

Q7. Mathematics Help Center and other walk-in help programs.

Students, especially preservice teachers, staff a Mathematics Help Center. The student helpers seem to like working in the Center.

Q8. Level of innovation in curriculum and teaching methods.

Inquiry style of teaching made several regular, as well as special topic, courses quite innovative.

Q9. Ways to recruit students to become mathematics majors, through freshman mathematics courses, extracurricular means, recruitment materials.

The large number of UNH mathematics graduates now teaching in schools seem to be the primary extra in recruitment at UNH.

Q10. Support programs for women and/or under represented groups.

Nothing special.

Q11. Use of technology.

Well equipped workstation lab and also PC labs (acquired on ILI grant).

Q13. Facilities for mathematics majors: departmental computing lab, study room, mathematics library.

Beyond the good computing facilities, I did not see any other facilities for students. No in-house library.

Q14, 15, 16, 17. Summer Research Opportunities, Independent study, Honors courses, Special Topics Courses, Colloquium for students.

Several special topics courses; some honor sections and little else available.

Q19. Training programs for junior faculty, particularly for non-native speaking faculty.

The department has a rich atmosphere of interest in teaching that leads to lots of informal help for new faculty. Junior faculty are welcome to participate in seminar and workshop activities supported by a FIPSE TA development grant to the department.

Contact: Joan Ferrini-Mundy, j_ferrini@unhh.unh.edu